INTERNATIONAL MATHEMATICAL OLYMPIADS 1959-1977

NEW MATHEMATICAL LIBRARY

PUBLISHED BY

THE MATHEMATICAL ASSOCIATION OF AMERICA

The New Mathematical Library (NML) was begun in 1961 by the School Mathematics Study Group to make available to high school students short expository books on various topics not usually covered in the high school syllabus. In a decade the NML matured into a steadily growing series of some twenty titles of interest not only to the originally intended audience, but to college students and teachers at all levels. Previously published by Random House and L. W. Singer, the NML became a publication series of the Mathematical Association of America (MAA) in 1975. Under the auspices of the MAA the NML will continue to grow and will remain dedicated to its original and expanded purposes.

INTERNATIONAL MATHEMATICAL OLYMPIADS 1959-1977

Compiled and with solutions by

Samuel L. Greitzer

Rutgers University

27

THE MATHEMATICAL ASSOCIATION
OF AMERICA

Drawings by Buehler & McFadden

Fourth Printing
 Published in Washington, D.C. by the
Mathematical Association of America

Library of Congress Catalog Card Number: 78-54027

Complete Set ISBN-0-88385-600-X

Vol. 27 0-88385-627-1

Manufactured in the United States of America

To my wife

Ethel

In loving memory

NEW MATHEMATICAL LIBRARY

1 Numbers: Rational and Irrational *by Ivan Niven*
2 What is Calculus About? *by W. W. Sawyer*
3 An Introduction to Inequalities *by E. F. Beckenbach and R. Bellman*
4 Geometric Inequalities *by N. D. Kazarinoff*
5 The Contest Problem Book I Annual High School Mathematics Examinations 1950–1960. Compiled and with solutions *by Charles T. Salkind*
6 The Lore of Large Numbers *by P. J. Davis*
7 Uses of Infinity *by Leo Zippin*
8 Geometric Transformations I *by I. M. Yaglom, translated by A. Shields*
9 Continued Fractions *by Carl D. Olds*
10 Graphs and Their Uses *by Oystein Ore*
11, 12 Hungarian Problem Books I and II, Based on the Eötvös Competitions 1894–1905 and 1906–1928, *translated by E. Rapaport*
13 Episodes from the Early History of Mathematics *by A. Aaboe*
14 Groups and Their Graphs *by I. Grossman and W. Magnus*
15 The Mathematics of Choice *by Ivan Niven*
16 From Pythagoras to Einstein *by K. O. Friedrichs*
17 The Contest Problem Book II Annual High School Mathematics Examinations 1961–1965. Compiled and with solutions *by Charles T. Salkind*
18 First Concepts of Topology *by W. G. Chinn and N. E. Steenrod*
19 Geometry Revisited *by H. S. M. Coxeter and S. L. Greitzer*
20 Invitation to Number Theory *by Oystein Ore*
21 Geometric Transformations II *by I. M. Yaglom, translated by A. Shields*
22 Elementary Cryptanalysis—A Mathematical Approach *by A. Sinkov*
23 Ingenuity in Mathematics *by Ross Honsberger*
24 Geometric Transformations III *by I. M. Yaglom, translated by A. Shenitzer*
25 The Contest Problem Book III Annual High School Mathematics Examinations 1966–1972. Compiled and with solutions *by C. T. Salkind and J. M. Earl*
26 Mathematical Methods in Science *by George Pólya*
27 International Mathematical Olympiads—1959–1977. Compiled and with solutions *by S. L. Greitzer*
28 The Mathematics of Games and Gambling *by Edward W. Packel*
29 The Contest Problem Book IV Annual High School Mathematics Examinations 1973–1982. Compiled and with solutions *by R. A. Artino, A. M. Gaglione and N. Shell*
30 The Role of Mathematics in Science *by M. M. Schiffer and L. Bowden*

Other titles in preparation

Editors' Note

The Mathematical Association of America is pleased to add the International Mathematical Olympiad Contests, 1959-1977, to the distinguished problem collections published in the New Mathematical Library. The basic text was prepared by S. L. Greitzer. The educational impact of such problems in stimulating mathematical thinking of young students and its long range effects have been eloquently described both from the viewpoint of the participant and that of the mature mathematician in retrospect by Gabor Szegö in his preface to the Hungarian Problem Books, volumes 11 and 12 of this NML series.

Our aim in the present collection is not only to help the high school student satisfy his curiosity by presenting solutions with tools familiar to him, but also to instruct him in the use of more sophisticated methods and different modes of attack by including explanatory material and alternate solutions. For problem solvers each problem is a challenging entity to be conquered; for theory spinners, each problem is the proof of their pudding. It is the fruitful synthesis of these seemingly antithetical forces that we have tried to achieve.

We are extremely grateful to Samuel L. Greitzer, the ingenious problem solver and devoted coach who helped lead the U. S. Olympiad team to victory in 1977, for having compiled the bulk of the solutions; some of them are based on the contestants' papers. We also acknowledge gratefully the many alternate solutions and elaborations contributed by Peter Ungar.

The editors of the present collection have occasionally departed somewhat from the wording of the problems originally presented to the English-speaking contestants. This was done in the interest of clarity and smooth style; since translations from one language into another are seldom completely faithful, we felt that such small departures were permissible.

We close this foreword by quoting G. Szegö's concluding observation from his preface to NML volumes 11 and 12:

> "We should not forget that the solution of any worthwhile problem very rarely comes to us easily and without hard work; it is rather the result of intellectual effort of days or weeks or months. Why should the

young mind be willing to make this supreme effort? The explanation is probably the instinctive preference for certain values, that is, the attitude which rates intellectual effort and spiritual achievement higher than material advantage. Such a valuation can be only the result of a long cultural development of environment and public spirit which is difficult to accelerate by governmental aid or even by more intensive training in mathematics. The most effective means may consist of transmitting to the young mind the beauty of intellectual work and the feeling of satisfaction following a great and successful mental effort. The hope is justified that the present book might aid exactly in this respect and that it represents a good step in the right direction."

Basil Gordon
William G. Chinn
Ivan Niven
Max Schiffer
Anneli Lax

December, 1977

Preface

This volume is a collection of all the problems in the International Mathematical Olympiads (IMO) from the First (1959) through the Nineteenth (1977) together with their solutions.

To explain how the problems are selected and the contests administered, I give a bit of the historical background.

Various countries have conducted national mathematical contests for a long time. The Hungarian Eötvös Competition (begun in 1894, see NML vols. 11 and 12) is a famous example. In 1959 Rumania invited Hungary, Bulgaria, Poland, Czechoslovakia, the German Democratic Republic (G.D.R.) and the U.S.S.R. to participate in the First I.M.O. After a slow start the number of participating nations grew. Finland joined in 1965, Great Britain, France and Italy in 1967; and since then the number of participating nations grew rapidly, reaching twenty-one by 1977.

The U.S.A. first participated in the IMO in 1974. Travel to this competition in Erfurt, G.D.R., was made possible by a generous grant by the Spencer Foundation. Also, the National Science Foundation funded a three-week training session at Rutgers University for the American team prior to its departure. Preparatory work started in 1972 when the Subcommittee on the U.S.A. Mathematical Olympiad of the Mathematical Association of America's Committee on High School Contests organized the first U.S.A. Mathematical Olympiad (USAMO). This contest examination was written by Murray Klamkin of the University of Alberta, Alberta, Canada and administered by this writer to the top 100 scorers (out of 300,000) on the Annual High School Mathematics Examination.

In 1975, the training session was again held at Rutgers University and funded by NSF. The 1975 IMO was held in Burgas, Bulgaria, and travel was made possible by grants from Johnson and Johnson Foundation, Minnesota Mining and Manufacturing Corporation, the Spencer Foundation, Standard Oil of California, and Xerox Corporation.

The 1976 training session was held at the U.S. Naval Academy and the 1977 Training Session was held at the U.S. Military Academy. These training sessions were funded by grants from the Army Research Office and the Office of Naval Research. Travel to the 1976 IMO in Lienz, Austria, and the 1977 IMO in Belgrade, Yugoslavia, was made possible by grants from the Army Research Office.

Each nation competing in the IMO generally sends a team of eight students, a delegate and a deputy. Delegates are requested to send to the host nation a maximum of five problems that they deem suitable for the IMO; the host nation picks about 18 problems, and the delegates meet at the beginning of each session to select the six problems to be solved. These are then translated into the four official languages, English, French, German and Russian, and prepared for the contestants.

On the first day there are impressive opening ceremonies, and the first three problems are given to the teams; they work on the remaining problems on the second day, while the Jury, i.e., the delegates and deputies from each country, begins grading the papers of its own team. The host country provides a coordinating team to check the papers; changes of grades are carefully discussed. The grades are made final and the Jury selects recipients of prizes for high scores and for especially elegant solutions. In the meantime the students are free to participate in excursions and other activities planned by the host nation. After a crowded two weeks, there are closing ceremonies where prizes are awarded, usually followed by a farewell party.

I have thoroughly enjoyed working with the students. They were exemplary representatives of the U.S. and I believe that they will become fine mathematicians. One—Gerhard Arenstorf—was tragically killed in an accident in August, 1974.

My thanks go to Murray Klamkin for sharing the joys and burdens of coaching the teams. I also gratefully acknowledge the work of David P. Roselle, Secretary of the Mathematical Association of America, for helping to make arrangements for the training sessions and for travel to Europe. All the editors of the New Mathematical Library are to be commended for their care and patience in going over the manuscript; my thanks especially to Anneli Lax, Basil Gordon and Ivan Niven. Thanks are due also to Peter Ungar for many improvements and additions. These colleagues were evidently fascinated by these problems and often added their own solutions to the text. I hope the reader will be equally fascinated and will enjoy solving the problems as much as we all did.

Samuel L. Greitzer
Rutgers University

Contents

Editors' Note vii

Preface ix

Problems 1

Solutions 21

Olympiad 1, 1959 21
Olympiad 2,1960 27
Olympiad 3, 1961 35
Olympiad 4, 1962 45
Olympiad 5, 1963 56
Olympiad 6, 1964 64
Olympiad 7, 1965 71
Olympiad 8, 1966 82
Olympiad 9, 1967 97
Olympiad 10, 1968 106
Olympiad 11, 1969 112
Olympiad 12, 1970 121
Olympiad 13, 1971 131
Olympiad 14, 1972 141
Olympiad 15, 1973 148
Olympiad 16, 1974 159
Olympiad 17, 1975 167
Olympiad 18, 1976 178
Olympiad 19, 1977 185

List of Symbols 195

Glossary 196

References 203

Problems

First International Olympiad, 1959

1959/1. Prove that the fraction $\dfrac{21n + 4}{14n + 3}$ is irreducible for every natural number n.

1959/2. For what real values of x is

$$\sqrt{(x + \sqrt{2x - 1})} + \sqrt{(x - \sqrt{2x - 1})} = A\ ,$$

given (a) $A = \sqrt{2}$, (b) $A = 1$, (c) $A = 2$, where only non-negative real numbers are admitted for square roots?

1959/3. Let a, b, c be real numbers. Consider the quadratic equation in $\cos x$:

$$a \cos^2 x + b \cos x + c = 0\ .$$

Using the numbers $a, b, c,$ form a quadratic equation in $\cos 2x$ whose roots are the same as those of the original equation. Compare the equations in $\cos x$ and $\cos 2x$ for $a = 4$, $b = 2$, $c = -1$.

1959/4. Construct a right triangle with given hypotenuse c such that the median drawn to the hypotenuse is the geometric mean of the two legs of the triangle.

1959/5. An arbitrary point M is selected in the interior of the segment AB. The squares $AMCD$ and $MBEF$ are constructed on the same side of AB, with the segments AM and MB as their respective bases. The circles circumscribed about these squares, with centers P and Q, intersect at M and also at another point N. Let N' denote the point of intersection of the straight lines AF and BC.

(a) Prove that the points N and N' coincide.

(b) Prove that the straight lines MN pass through a fixed point S independent of the choice of M.

(c) Find the locus of the midpoints of the segments PQ as M varies between A and B.

1959/6. Two planes, P and Q, intersect along the line p. The point A is given in the plane P, and the point C in the plane Q; neither of these points lies on the straight line p. Construct an isosceles trapezoid $ABCD$ (with AB parallel to CD) in which a circle can be inscribed, and with vertices B and D lying in the planes P and Q respectively.

Second International Olympiad, 1960

1960/1. Determine all three-digit numbers N having the property that N is divisible by 11, and $N/11$ is equal to the sum of the squares of the digits of N.

1960/2. For what values of the variable x does the following inequality hold:

$$\frac{4x^2}{(1-\sqrt{1+2x}\,)^2} < 2x + 9 \ ?$$

1960/3. In a given right triangle ABC, the hypotenuse BC, of length a, is divided into n equal parts (n an odd integer). Let α be the acute angle subtending, from A, that segment which contains the midpoint of the hypotenuse. Let h be the length of the altitude to the hypotenuse of the triangle. Prove:

$$\tan \alpha = \frac{4nh}{(n^2-1)a}\,.$$

1960/4. Construct triangle ABC, given h_a, h_b (the altitudes from A and B) and m_a, the median from vertex A.

1960/5. Consider the cube $ABCDA'B'C'D'$ (with face $ABCD$ directly above face $A'B'C'D'$).

(a) Find the locus of the midpoints of segments XY, where X is any point of AC and Y is any point of $B'D'$.

(b) Find the locus of points Z which lie on the segments XY of part (a) with $ZY = 2XZ$.

1960/6. Consider a cone of revolution with an inscribed sphere tangent to the base of the cone. A cylinder is circumscribed about this sphere so that one of its bases lies in the base of the cone. Let V_1 be the volume of the cone and V_2 the volume of the cylinder.

(a) Prove that $V_1 \neq V_2$.

(b) Find the smallest number k for which $V_1 = kV_2$; for this case, construct the angle subtended by a diameter of the base of the cone at the vertex of the cone.

1960/7. An isosceles trapezoid with bases a and c and altitude h is given.

(a) On the axis of symmetry of this trapezoid, find all points P such that both legs of the trapezoid subtend right angles at P.

(b) Calculate the distance of P from either base.

(c) Determine under what conditions such points P actually exist. (Discuss various cases that might arise.)

Third International Olympiad, 1961

1961/1. Solve the system of equations:

$$x + y + z = a$$
$$x^2 + y^2 + z^2 = b^2$$
$$xy = z^2$$

where a and b are constants. Give the conditions that a and b must satisfy so that x, y, z (the solutions of the system) are distinct positive numbers.

1961/2. Let a, b, c be the sides of a triangle, and T its area. Prove: $a^2 + b^2 + c^2 \geqslant 4\sqrt{3}\,T$. In what case does equality hold?

1961/3. Solve the equation $\cos^n x - \sin^n x = 1$, where n is a natural number.

1961/4. Consider triangle $P_1P_2P_3$ and a point P within the triangle. Lines P_1P, P_2P, P_3P intersect the opposite sides in points Q_1, Q_2, Q_3 respectively. Prove that, of the numbers

$$\frac{P_1P}{PQ_1}, \quad \frac{P_2P}{PQ_2}, \quad \frac{P_3P}{PQ_3},$$

at least one is $\leqslant 2$ and at least one is $\geqslant 2$.

1961/5. Construct triangle ABC if $AC = b, AB = c$ and $\angle AMB = \omega$, where M is the midpoint of segment BC and $\omega < 90°$. Prove that a solution exists if and only if

$$b \tan \frac{\omega}{2} \leqslant c < b .$$

In what case does the equality hold?

1961/6. Consider a plane ε and three non-collinear points A, B, C on the same side of ε; suppose the plane determined by these three points is not parallel to ε. In plane ε take three arbitrary points A', B', C'. Let L, M, N be the midpoints of segments AA', BB', CC'; let G be the centroid of triangle LMN. (We will not consider positions of the points A', B', C' such that the points L, M, N do not form a triangle.†) What is the locus of point G as A', B', C' range independently over the plane ε?

Fourth International Olympiad, 1962

1962/1. Find the smallest natural number n which has the following properties:

(a) Its decimal representation has 6 as the last digit.

(b) If the last digit 6 is erased and placed in front of the remaining digits, the resulting number is four times as large as the original number n.

1962/2. Determine all real numbers x which satisfy the inequality:

$$\sqrt{3 - x} - \sqrt{x + 1} > \frac{1}{2} .$$

1962/3. Consider the cube $ABCDA'B'C'D'$ ($ABCD$ and $A'B'C'D'$ are the upper and lower bases, respectively, and edges AA', BB', CC', DD' are parallel). The point X moves at constant speed along the perimeter of the square $ABCD$ in the direction $ABCDA$, and the point Y moves at the same rate along the perimeter of the square $B'C'CB$ in the direction $B'C'CBB'$. Points X and Y begin their motion at the same instant from the starting positions A and B', respectively. Determine and draw the locus of the midpoints of the segments XY.

1962/4. Solve the equation $\cos^2 x + \cos^2 2x + \cos^2 3x = 1$.

1962/5. On the circle K there are given three distinct points A, B, C. Construct (using only straightedge and compasses) a fourth point D on K such that a circle can be inscribed in the quadrilateral thus obtained.

†Since the centroid of any three points can be defined (even if the points are collinear, or if two or all three coincide), this problem can be solved without the restriction in parentheses.

1962/6. Consider an isosceles triangle. Let r be the radius of its circumscribed circle and ρ the radius of its inscribed circle. Prove that the distance d between the centers of these two circles is

$$d = \sqrt{r(r-2\rho)} \ .$$

1962/7. The tetrahedron $SABC$ has the following property: there exist five spheres, each tangent to the edges $SA, SB, SC, BC, CA, AB,$ or to their extensions.

(a) Prove that the tetrahedron $SABC$ is regular.

(b) Prove conversely that for every regular tetrahedron five such spheres exist.

Fifth International Olympiad, 1963

1963/1. Find all real roots of the equation

$$\sqrt{x^2 - p} + 2\sqrt{x^2 - 1} = x \ ,$$

where p is a real parameter.

1963/2. Point A and segment BC are given. Determine the locus of points in space which are vertices of right angles with one side passing through A, and the other side intersecting the segment BC.

1963/3. In an n-gon all of whose interior angles are equal, the lengths of consecutive sides satisfy the relation

$$a_1 \geqslant a_2 \geqslant \cdots \geqslant a_n .$$

Prove that $a_1 = a_2 = \cdots = a_n$.

1963/4. Find all solutions x_1, x_2, x_3, x_4, x_5 of the system

$$x_5 + x_2 = yx_1 \tag{1}$$

$$x_1 + x_3 = yx_2 \tag{2}$$

$$x_2 + x_4 = yx_3 \tag{3}$$

$$x_3 + x_5 = yx_4 \tag{4}$$

$$x_4 + x_1 = yx_5 \ , \tag{5}$$

where y is a parameter.

1963/5. Prove that $\cos\frac{\pi}{7} - \cos\frac{2\pi}{7} + \cos\frac{3\pi}{7} = \frac{1}{2}$

1963/6. Five students, A, B, C, D, E, took part in a contest. One prediction was that the contestants would finish in the order ABCDE. This prediction was very poor. In fact no contestant finished in the position predicted, and no two contestants predicted to finish consecutively† actually did so. A second prediction had the contestants finishing in the order DAECB. This prediction was better. Exactly two of the contestants finished in the places predicted, and two disjoint pairs of students predicted to finish consecutively actually did so. Determine the order in which the contestants finished.

Sixth International Olympiad, 1964

1964/1. (a) Find all positive integers n for which $2^n - 1$ is divisible by 7.

(b) Prove that there is no positive integer n for which $2^n + 1$ is divisible by 7.

1964/2. Suppose a, b, c are the sides of a triangle. Prove that

$$a^2(b + c - a) + b^2(c + a - b) + c^2(a + b - c) \leqslant 3abc \; .$$

1964/3. A circle is inscribed in triangle ABC with sides a, b, c. Tangents to the circle parallel to the sides of the triangle are constructed. Each of these tangents cuts off a triangle from $\triangle ABC$. In each of these triangles, a circle is inscribed. Find the sum of the areas of all four inscribed circles (in terms of a, b, c).

1964/4. Seventeen people correspond by mail with one another—each one with all the rest. In their letters only three different topics are discussed. Each pair of correspondents deals with only one of these topics. Prove that there are at least three people who write to each other about the same topic.

1964/5. Suppose five points in a plane are situated so that no two of the straight lines joining them are parallel, perpendicular, or coincident. From each point perpendiculars are drawn to all the lines joining the other four points. Determine the maximum number of intersections that these perpendiculars can have.

1964/6. In tetrahedron $ABCD$, vertex D is connected with D_0, the centroid of $\triangle ABC$. Lines parallel to DD_0 are drawn through A, B and

†Two students X and Y are "predicted to finish consecutively" means here that they are predicted to finish in the *order* XY with no contestant finishing between them.

C. These lines intersect the planes BCD, CAD and ABD in points A_1, B_1 and C_1, respectively. Prove that the volume of $ABCD$ is one third the volume of $A_1B_1C_1D_0$. Is the result true if point D_0 is selected anywhere within $\triangle ABC$?

Seventh International Olympiad, 1965

1965/1. Determine all values x in the interval $0 \leqslant x \leqslant 2\pi$ which satisfy the inequality

$$2 \cos x \leqslant |\sqrt{1 + \sin 2x} - \sqrt{1 - \sin 2x}| \leqslant \sqrt{2}\ .$$

1965/2. Consider the system of equations

$$a_{11}x_1 + a_{12}x_2 + a_{13}x_3 = 0$$
$$a_{21}x_1 + a_{22}x_2 + a_{23}x_3 = 0$$
$$a_{31}x_1 + a_{32}x_2 + a_{33}x_3 = 0$$

with unknowns x_1, x_2, x_3. The coefficients satisfy the conditions:

(a) a_{11}, a_{22}, a_{33} are positive numbers;
(b) the remaining coefficients are negative numbers;
(c) in each equation, the sum of the coefficients is positive.

Prove that the given system has only the solution $x_1 = x_2 = x_3 = 0$.

1965/3. Given the tetrahedron $ABCD$ whose edges AB and CD have lengths a and b respectively. The distance between the skew lines AB and CD is d, and the angle between them is ω. Tetrahedron $ABCD$ is divided into two solids by plane ε, parallel to lines AB and CD. The ratio of the distances of ε from AB and CD is equal to k. Compute the ratio of the volumes of the two solids obtained.

1965/4. Find all sets of four real numbers x_1, x_2, x_3, x_4 such that the sum of any one and the product of the other three is equal to 2.

1965/5. Consider $\triangle OAB$ with acute angle AOB. Through a point $M \neq O$ perpendiculars are drawn to OA and OB, the feet of which are P and Q respectively. The point of intersection of the altitudes of $\triangle OPQ$ is H. What is the locus of H if M is permitted to range over (a) the side AB, (b) the interior of $\triangle OAB$?

1965/6. In a plane a set of n points ($n \geqslant 3$) is given. Each pair of points is connected by a segment. Let d be the length of the longest of these segments. We define a *diameter* of the set to be any connecting segment of length d. Prove that the number of diameters of the given set is at most n.

Eighth International Olympiad, 1966

1966/1. In a mathematical contest, three problems, A, B, C were posed. Among the participants there were 25 students who solved at least one problem each. Of all the contestants who did not solve problem A, the number who solved B was twice the number who solved C. The number of students who solved only problem A was one more than the number of students who solved A and at least one other problem. Of all students who solved just one problem, half did not solve problem A. How many students solved only problem B?

1966/2. Let a, b, c be the lengths of the sides of a triangle, and α, β, γ, respectively, the angles opposite these sides. Prove that if

$$a + b = \tan\frac{\gamma}{2}(a \tan\alpha + b \tan\beta),$$

the triangle is isosceles.

1966/3. Prove: The sum of the distances of the vertices of a regular tetrahedron from the center of its circumscribed sphere is less than the sum of the distances of these vertices from any other point in space.

1966/4. Prove that for every natural number n, and for every real number $x \neq k\pi/2^t$ $(t = 0, 1, \ldots, n;\ k$ any integer)

$$\frac{1}{\sin 2x} + \frac{1}{\sin 4x} + \cdots + \frac{1}{\sin 2^n x} = \cot x - \cot 2^n x .$$

1966/5. Solve the system of equations

$$\begin{aligned}
|a_1 - a_2|x_2 + |a_1 - a_3|x_3 + |a_1 - a_4|x_4 &= 1\\
|a_2 - a_1|x_1 + |a_2 - a_3|x_3 + |a_2 - a_4|x_4 &= 1\\
|a_3 - a_1|x_1 + |a_3 - a_2|x_2 + |a_3 - a_4|x_4 &= 1\\
|a_4 - a_1|x_1 + |a_4 - a_2|x_2 + |a_4 - a_3|x_3 &= 1
\end{aligned}$$

where a_1, a_2, a_3, a_4 are four different real numbers.

1966/6. In the interior of sides BC, CA, AB of triangle ABC, any points K, L, M, respectively, are selected. Prove that the area of at least one of the triangles AML, BKM, CLK is less than or equal to one quarter of the area of triangle ABC.

Ninth International Olympiad, 1967

1967/1. Let $ABCD$ be a parallelogram with side lengths $AB = a$, $AD = 1$, and with $\angle BAD = \alpha$. If $\triangle ABD$ is acute, prove that the four circles of radius 1 with centers A, B, C, D cover the parallelogram if and only if

$$a \leqslant \cos\alpha + \sqrt{3}\sin\alpha\ .$$

1967/2. Prove that if one and only one edge of a tetrahedron is greater than 1, then its volume is $\leqslant 1/8$.

1967/3. Let k, m, n be natural numbers such that $m + k + 1$ is a prime greater than $n + 1$. Let $c_s = s(s + 1)$. Prove that the product

$$(c_{m+1} - c_k)(c_{m+2} - c_k) \cdots (c_{m+n} - c_k)$$

is divisible by the product $c_1 c_2 \cdots c_n$.

1967/4. Let $A_0B_0C_0$ and $A_1B_1C_1$ be any two acute-angled triangles. Consider all triangles ABC that are similar to $\triangle A_1B_1C_1$ (so that vertices A_1, B_1, C_1 correspond to vertices $A, B, C,$ respectively) and circumscribed about triangle $A_0B_0C_0$ (where A_0 lies on BC, B_0 on CA, and C_0 on AB). Of all such possible triangles, determine the one with maximum area, and construct it.

1967/5. Consider the sequence $\{c_n\}$, where

$$c_1 = a_1 + a_2 + \cdots + a_8$$

$$c_2 = a_1^2 + a_2^2 + \cdots + a_8^2$$

$$\cdots\cdots\cdots\cdots\cdots\cdots$$

$$c_n = a_1^n + a_2^n + \cdots + a_8^n\ ,$$

$$\cdots\cdots\cdots\cdots\cdots\cdots$$

in which $a_1, a_2, \ldots, a_8$ are real numbers not all equal to zero. Suppose that an infinite number of terms of the sequence $\{c_n\}$ are equal to zero. Find all natural numbers n for which $c_n = 0$.

1967/6. In a sports contest, there were m medals awarded on n successive days $(n > 1)$. On the first day, one medal and $1/7$ of the remaining $m - 1$ medals were awarded. On the second day, two medals and $1/7$ of the now remaining medals were awarded; and so on. On the n-th and last day, the remaining n medals were awarded. How many days did the contest last, and how many medals were awarded altogether?

in which $a_1, a_2, \ldots, a_8$ are real numbers not all equal to zero. Suppose that an infinite number of terms of the sequence $\{c_n\}$ are equal to zero. Find all natural numbers n for which $c_n = 0$.

Tenth International Olympiad, 1968

1968/1. Prove that there is one and only one triangle whose side lengths are consecutive integers, and one of whose angles is twice as large as another.

1968/2. Find all natural numbers x such that the product of their digits (in decimal notation) is equal to $x^2 - 10x - 22$.

1968/3. Consider the system of equations

$$\begin{aligned} ax_1^2 + bx_1 + c &= x_2 \\ ax_2^2 + bx_2 + c &= x_3 \\ &\cdots\cdots \\ ax_{n-1}^2 + bx_{n-1} + c &= x_n \\ ax_n^2 + bx_n + c &= x_1, \end{aligned}$$

with unknowns $x_1, x_2, \ldots, x_n$, where a, b, c are real and $a \neq 0$. Let $\triangle = (b-1)^2 - 4ac$. Prove that for this system

(a) if $\triangle < 0$, there is no solution,
(b) if $\triangle = 0$, there is exactly one solution,
(c) if $\triangle > 0$, there is more than one solution.

1968/4. Prove that in every tetrahedron there is a vertex such that the three edges meeting there have lengths which are the sides of a triangle.

1968/5. Let f be a real-valued function defined for all real numbers x such that, for some positive constant a, the equation

$$f(x+a) = \frac{1}{2} + \sqrt{f(x) - [f(x)]^2}$$

holds for all x.

(a) Prove that the function f is periodic (i.e., there exists a positive number b such that $f(x+b) = f(x)$ for all x).

(b) For $a = 1$, give an example of a non-constant function with the required properties.

1968/6. For every natural number n, evaluate the sum

$$\sum_{k=0}^{\infty}\left[\frac{n+2^k}{2^{k+1}}\right]=\left[\frac{n+1}{2}\right]+\left[\frac{n+2}{4}\right]+\cdots+\left[\frac{n+2^k}{2^{k+1}}\right]+\cdots .$$

(The symbol $[x]$ denotes the greatest integer not exceeding x.)

Eleventh International Olympiad, 1969

1969/1. Prove that there are infinitely many natural numbers a with the following property: the number $z = n^4 + a$ is not prime for any natural number n.

1969/2. Let $a_1, a_2, \ldots, a_n$ be real constants, x a real variable, and

$$f(x) = \cos(a_1 + x) + \tfrac{1}{2}\cos(a_2 + x) + \tfrac{1}{4}\cos(a_3 + x) + \cdots + \frac{1}{2^{n-1}}\cos(a_n + x) .$$

Given that $f(x_1) = f(x_2) = 0$, prove that $x_2 - x_1 = m\pi$ for some integer m.

1969/3. For each value of $k = 1, 2, 3, 4, 5$, find necessary and sufficient conditions on the number $a > 0$ so that there exists a tetrahedron with k edges of length a, and the remaining $6 - k$ edges of length 1.

1969/4. A semicircular arc γ is drawn on AB as diameter. C is a point on γ other than A and B, and D is the foot of the perpendicular from C to AB. We consider three circles, $\gamma_1, \gamma_2, \gamma_3$, all tangent to the line AB. Of these, γ_1 is inscribed in $\triangle ABC$, while γ_2 and γ_3 are both tangent to CD and to γ, one on each side of CD. Prove that γ_1, γ_2 and γ_3 have a second tangent in common.

1969/5. Given $n > 4$ points in the plane such that no three are collinear. Prove that there are at least $\binom{n-3}{2}$ convex quadrilaterals whose vertices are four of the given points.

1969/6. Prove that for all real numbers $x_1, x_2, y_1, y_2, z_1, z_2$ with $x_1 > 0$, $x_2 > 0$, $x_1y_1 - z_1^2 > 0$, $x_2y_2 - z_2^2 > 0$, the inequality

$$\frac{8}{(x_1 + x_2)(y_1 + y_2) - (z_1 + z_2)^2} \leqslant \frac{1}{x_1y_1 - z_1^2} + \frac{1}{x_2y_2 - z_2^2}$$

is satisfied. Give necessary and sufficient conditions for equality.

Twelfth International Olympiad, 1970

1970/1. Let M be a point on the side AB of $\triangle ABC$. Let r_1, r_2 and r be the radii of the inscribed circles of triangles AMC, BMC and ABC. Let q_1, q_2 and q be the radii of the escribed circles of the same triangles that lie in the angle ACB. Prove that

$$\frac{r_1}{q_1} \cdot \frac{r_2}{q_2} = \frac{r}{q} .$$

1970/2. Let a, b and n be integers greater than 1, and let a and b be the bases of two number systems. A_{n-1} and A_n are numbers in the system with base a, and B_{n-1} and B_n are numbers in the system with base b; these are related as follows:

$$\left.\begin{array}{ll} A_n = x_n x_{n-1} \cdots x_0, & A_{n-1} = x_{n-1} x_{n-2} \cdots x_0 , \\ B_n = x_n x_{n-1} \cdots x_0, & B_{n-1} = x_{n-1} x_{n-2} \cdots x_0 , \end{array}\right\} (x_n \neq 0, x_{n-1} \neq 0) .$$

Prove:

$$\frac{A_{n-1}}{A_n} < \frac{B_{n-1}}{B_n} \text{ if and only if } a > b .$$

1970/3. The real numbers $a_0, a_1, \ldots, a_n, \ldots$ satisfy the condition:

$$1 = a_0 \leqslant a_1 \leqslant a_2 \leqslant \cdots \leqslant a_n \leqslant \cdots .$$

The numbers $b_1, b_2, \ldots, b_n, \ldots$ are defined by

$$b_n = \sum_{k=1}^{n} \left(1 - \frac{a_{k-1}}{a_k}\right) \frac{1}{\sqrt{a_k}} .$$

(a) Prove that $0 \leqslant b_n < 2$ for all n.

(b) Given c with $0 \leqslant c < 2$, prove that there exist numbers $a_0, a_1, \ldots$ with the above properties such that $b_n > c$ for large enough n.

1970/4. Find the set of all positive integers n with the property that the set $\{n, n+1, n+2, n+3, n+4, n+5\}$ can be partitioned into two sets such that the product of the numbers in one set equals the product of the numbers in the other set.

1970/5. In the tetrahedron $ABCD$, angle BDC is a right angle. Suppose that the foot H of the perpendicular from D to the plane ABC is the intersection of the altitudes of $\triangle ABC$. Prove that

$$(AB + BC + CA)^2 \leqslant 6(AD^2 + BD^2 + CD^2) .$$

For what tetrahedra does equality hold?

1970/6. In a plane there are 100 points, no three of which are collinear. Consider all possible triangles having these points as vertices. Prove that no more than 70% of these triangles are acute-angled.

Thirteenth International Olympiad, 1971

1971/1. Prove that the following assertion is true for $n = 3$ and $n = 5$, and that it is false for every other natural number $n > 2$:

If $a_1, a_2, \ldots, a_n$ are arbitrary real numbers, then

$$(a_1 - a_2)(a_1 - a_3) \cdots (a_1 - a_n) + (a_2 - a_1)(a_2 - a_3) \cdots (a_2 - a_n) + \cdots + (a_n - a_1)(a_n - a_2) \cdots (a_n - a_{n-1}) \geqslant 0 .$$

1971/2. Consider a convex polyhedron P_1 with nine vertices $A_1, A_2, \ldots, A_9$; let P_i be the polyhedron obtained from P_1 by a translation that moves vertex A_1 to A_i $(i = 2, 3, \ldots, 9)$. Prove that at least two of the polyhedra $P_1, P_2, \ldots, P_9$ have an interior point in common.

1971/3. Prove that the set of integers of the form $2^k - 3$ $(k = 2, 3, \ldots)$ contains an infinite subset in which every two members are relatively prime.

1971/4. All the faces of tetrahedron $ABCD$ are acute-angled triangles. We consider all closed polygonal paths of the form $XYZTX$ defined as follows: X is a point on edge AB distinct from A and B; similarly, Y, Z, T are interior points of edges $BC, CD, DA,$ respectively. Prove:

(a) If $\angle DAB + \angle BCD \neq \angle CDA + \angle ABC$, then among the polygonal paths, there is none of minimal length.

(b) If $\angle DAB + \angle BCD = \angle CDA + \angle ABC$, then there are infinitely many shortest polygonal paths, their common length being $2AC \sin(\alpha/2)$, where $\alpha = \angle BAC + \angle CAD + \angle DAB$.

1971/5. Prove that for every natural number m, there exists a finite set S of points in a plane with the following property: For every point A in S, there are exactly m points in S which are at unit distance from A.

1971/6. Let $A = (a_{ij})$ $(i, j = 1, 2, \ldots, n)$ be a square matrix whose elements are non-negative integers. Suppose that whenever an element $a_{ij} = 0$, the sum of the elements in the ith row and the jth column is $\geqslant n$. Prove that the sum of all the elements of the matrix is $\geqslant n^2/2$.

Fourteenth International Olympiad, 1972

1972/1. Prove that from a set of ten distinct two-digit numbers (in the decimal system), it is possible to select two disjoint subsets whose members have the same sum.

1972/2. Prove that if $n \geqslant 4$, every quadrilateral that can be inscribed in a circle can be dissected into n quadrilaterals each of which is inscribable in a circle.

1972/3. Let m and n be arbitrary non-negative integers. Prove that

$$\frac{(2m)!\,(2n)!}{m!n!\,(m+n)!}$$

is an integer. $(0! = 1.)$

1972/4. Find all solutions $(x_1, x_2, x_3, x_4, x_5)$ of the system of inequalities

$$\begin{aligned}
(x_1^2 - x_3x_5)(x_2^2 - x_3x_5) &\leqslant 0\\
(x_2^2 - x_4x_1)(x_3^2 - x_4x_1) &\leqslant 0\\
(x_3^2 - x_5x_2)(x_4^2 - x_5x_2) &\leqslant 0\\
(x_4^2 - x_1x_3)(x_5^2 - x_1x_3) &\leqslant 0\\
(x_5^2 - x_2x_4)(x_1^2 - x_2x_4) &\leqslant 0
\end{aligned}$$

where x_1, x_2, x_3, x_4, x_5 are positive real numbers.

1972/5. Let f and g be real-valued functions defined for all real values of x and y, and satisfying the equation

$$f(x+y) + f(x-y) = 2f(x)g(y)$$

for all x, y. Prove that if $f(x)$ is not identically zero, and if $|f(x)| \leqslant 1$ for all x, then $|g(y)| \leqslant 1$ for all y.

1972/6. Given four distinct parallel planes, prove that there exists a regular tetrahedron with a vertex on each plane.

Fifteenth International Olympiad, 1973

1973/1. Point O lies on line g; $\overrightarrow{OP_1}, \overrightarrow{OP_2}, \ldots \overrightarrow{OP_n}$ are unit vectors such that points $P_1, P_2, \ldots, P_n$ all lie in a plane containing g and on one side of g. Prove that if n is odd,

$$|\overrightarrow{OP_1} + \overrightarrow{OP_2} + \cdots + \overrightarrow{OP_n}| \geqslant 1 .$$

Here $|\overrightarrow{OM}|$ denotes the length of vector $\overrightarrow{OM}$.

1973/2. Determine whether or not there exists a finite set M of points in space not lying in the same plane such that, for any two points A and B of M, one can select two other points C and D of M so that lines AB and CD are parallel and not coincident.

1973/3. Let a and b be real numbers for which the equation

$$x^4 + ax^3 + bx^2 + ax + 1 = 0$$

has at least one real solution. For all such pairs (a, b), find the minimum value of $a^2 + b^2$.

1973/4. A soldier needs to check on the presence of mines in a region having the shape of an equilateral triangle. The radius of action of his detector is equal to half the altitude of the triangle. The soldier leaves from one vertex of the triangle. What path should he follow in order to travel the least possible distance and still accomplish his mission?

1973/5. G is a set of non-constant functions of the real variable x of the form

$$f(x) = ax + b, \qquad a \text{ and } b \text{ are real numbers },$$

and G has the following properties:

(a) If f and g are in G, then $g \circ f$ is in G; here $(g \circ f)(x) = g[f(x)]$.

(b) If f is in G, then its inverse f^{-1} is in G; here the inverse of $f(x) = ax + b$ is $f^{-1}(x) = (x - b)/a$.

(c) For every f in G, there exists a real number x_f such that $f(x_f) = x_f$.

Prove that there exists a real number k such that $f(k) = k$ for all f in G.

1973/6. Let $a_1, a_2, \ldots, a_n$ be n positive numbers, and let q be a given real number such that $0 < q < 1$. Find n numbers $b_1, b_2, \ldots, b_n$ for which

(a) $a_k < b_k \quad$ for $k = 1, 2, \ldots, n$,

(b) $q < \dfrac{b_{k+1}}{b_k} < \dfrac{1}{q} \quad$ for $k = 1, 2, \ldots, n - 1$,

(c) $b_1 + b_2 + \cdots + b_n < \dfrac{1+q}{1-q}(a_1 + a_2 + \cdots + a_n)$.

Sixteenth International Olympiad, 1974

1974/1. Three players A, B and C play the following game: On each of three cards an integer is written. These three numbers p, q, r satisfy $0 < p < q < r$. The three cards are shuffled and one is dealt to each player. Each then receives the number of counters indicated by the card he holds. Then the cards are shuffled again; the counters remain with the players.

This process (shuffling, dealing, giving out counters) takes place for at least two rounds. After the last round, A has 20 counters in all, B has 10 and C has 9. At the last round B received r counters. Who received q counters on the first round?

1974/2. In the triangle ABC, prove that there is a point D on side AB such that CD is the geometric mean of AD and DB if and only if

$$\sin A \sin B \leqslant \sin^2 \frac{C}{2} .$$

1974/3. Prove that the number $\sum_{k=0}^{n}\binom{2n+1}{2k+1}2^{3k}$ is not divisible by 5 for any integer $n \geqslant 0$.

1974/4. Consider decompositions of an 8×8 chessboard into p non-overlapping rectangles subject to the following conditions:

(i) Each rectangle has as many white squares as black squares.

(ii) If a_i is the number of white squares in the i-th rectangle, then $a_1 < a_2 < \cdots < a_p$.

Find the maximum value of p for which such a decomposition is possible. For this value of p, determine all possible sequences $a_1, a_2, \ldots, a_p$.

1974/5. Determine all possible values of

$$S = \frac{a}{a+b+d} + \frac{b}{a+b+c} + \frac{c}{b+c+d} + \frac{d}{a+c+d}$$

where a, b, c, d are arbitrary positive numbers.

1974/6. Let P be a non-constant polynomial with integer coefficients. If $n(P)$ is the number of distinct integers k such that $(P(k))^2 = 1$, prove that $n(P) - \deg(P) \leqslant 2$, where $\deg(P)$ denotes the degree of the polynomial P.

Seventeenth International Olympiad, 1975

1975/1. Let x_i, y_i $(i = 1, 2, \ldots, n)$ be real numbers such that

$$x_1 \geqslant x_2 \geqslant \cdots \geqslant x_n \quad \text{and} \quad y_1 \geqslant y_2 \geqslant \cdots \geqslant y_n .$$

Prove that, if $z_1, z_2, \ldots, z_n$ is any permutation of $y_1, y_2, \ldots, y_n$, then

$$\sum_{i=1}^{n} (x_i - y_i)^2 \leqslant \sum_{i=1}^{n} (x_i - z_i)^2 .$$

1975/2. Let $a_1, a_2, a_3, \cdots$ be an infinite increasing sequence of positive integers. Prove that for every $p \geqslant 1$ there are infinitely many a_m which can be written in the form

$$a_m = xa_p + ya_q$$

with x, y positive integers and $q > p$.

1975/3. On the sides of an arbitrary triangle ABC, triangles ABR, BCP, CAQ are constructed externally with $\angle CBP = \angle CAQ = 45°$, $\angle BCP = \angle ACQ = 30°$, $\angle ABR = \angle BAR = 15°$. Prove that $\angle QRP = 90°$ and $QR = RP$.

1975/4. When 4444^{4444} is written in decimal notation, the sum of its digits is A. Let B be the sum of the digits of A. Find the sum of the digits of B. (A and B are written in decimal notation.)

1975/5. Determine, with proof, whether or not one can find 1975 points on the circumference of a circle with unit radius such that the distance between any two of them is a rational number.

1975/6. Find all polynomials P, in two variables, with the following properties:

(i) for a positive integer n and all real t, x, y

$$P(tx, ty) = t^n P(x, y)$$

(that is, P is homogeneous of degree n),

(ii) for all real a, b, c,

$$P(b + c, a) + P(c + a, b) + P(a + b, c) = 0 ,$$

(iii) $P(1, 0) = 1$.

Eighteenth International Olympiad, 1976

1976/1. In a plane convex quadrilateral of area 32, the sum of the lengths of two opposite sides and one diagonal is 16. Determine all possible lengths of the other diagonal.

1976/2. Let $P_1(x) = x^2 - 2$ and $P_j(x) = P_1(P_{j-1}(x))$ for $j = 2, 3, \cdots$. Show that, for any positive integer n, the roots of the equation $P_n(x) = x$ are real and distinct.

1976/3. A rectangular box can be filled completely with unit cubes. If one places as many cubes as possible, each with volume 2, in the box, so that their edges are parallel to the edges of the box, one can fill exactly 40% of the box. Determine the possible dimensions of all such boxes.

1976/4. Determine, with proof, the largest number which is the product of positive integers whose sum is 1976.

1976/5. Consider the system of p equations in $q = 2p$ unknowns $x_1, x_2, \ldots, x_q$:

$$\begin{aligned} a_{11}x_1 + a_{12}x_2 + \cdots \qquad & + a_{1q}x_q = 0 \\ a_{21}x_1 + a_{22}x_2 + \cdots \qquad & + a_{2q}x_q = 0 \\ \cdots\cdots\cdots\cdots\cdots & \cdots\cdots\cdots \\ a_{p1}x_1 + a_{p2}x_2 + \cdots \qquad & + a_{pq}x_q = 0 \end{aligned}$$

with every coefficient a_{ij} a member of the set $\{-1, 0, 1\}$. Prove that the system has a solution $(x_1, x_2, \ldots, x_q)$ such that

(a) all x_j $(j = 1, 2, \ldots, q)$ are integers,
(b) there is at least one value of j for which $x_j \neq 0$,
(c) $|x_j| \leqslant q$ $(j = 1, 2, \ldots, q)$.

1976/6. A sequence $\{u_n\}$ is defined by

$$u_0 = 2, \quad u_1 = 5/2, \quad u_{n+1} = u_n(u_{n-1}^2 - 2) - u_1 \quad \text{for} \quad n = 1, 2, \ldots.$$

Prove that for positive integers n,

$$[u_n] = 2^{[2^n - (-1)^n]/3},$$

where $[x]$ denotes the greatest integer $\leqslant x$.

Nineteenth International Mathematical Olympiad, 1977

1977/1. Equilateral triangles ABK, BCL, CDM, DAN are constructed inside the square $ABCD$. Prove that the midpoints of the four segments KL, LM, MN, NK and the midpoints of the eight segments AK, BK, BL, CL, CM, DM, DN, AN are the twelve vertices of a regular dodecagon.

1977/2. In a finite sequence of real numbers the sum of any seven successive terms is negative, and the sum of any eleven successive terms is positive. Determine the maximum number of terms in the sequence.

1977/3. Let n be a given integer >2, and let V_n be the set of integers $1+kn$, where $k=1,2,\ldots$. A number $m \in V_n$ is called *indecomposable in* V_n if there do not exist numbers $p, q \in V_n$ such that $pq = m$. Prove that there exists a number $r \in V_n$ that can be expressed as the product of elements indecomposable in V_n in more than one way. (Products which differ only in the order of their factors will be considered the same.)

1977/4. Four real constants a, b, A, B are given, and

$$f(\theta) = 1 - a\cos\theta - b\sin\theta - A\cos 2\theta - B\sin 2\theta\ .$$

Prove that if $f(\theta) \geqslant 0$ for all real θ, then

$$a^2 + b^2 \leqslant 2 \qquad \text{and} \qquad A^2 + B^2 \leqslant 1\ .$$

1977/5. Let a and b be positive integers. When $a^2 + b^2$ is divided by $a + b$, the quotient is q and the remainder is r. Find all pairs (a, b) such that $q^2 + r = 1977$.

1977/6. Let $f(n)$ be a function defined on the set of all positive integers and having all its values in the same set. Prove that if

$$f(n+1) > f(f(n))$$

for each positive integer n, then

$$f(n) = n \qquad \text{for each} \quad n\ .$$

Solutions

First International Olympiad, 1959

1959/1 First solution. We shall show that if $g > 0$ is a common factor of the numerator and denominator, then $g = 1$. Set

$$21n + 4 = gA \ , \qquad 14n + 3 = gB \ .$$

Then

$$g(3B - 2A) = 3gB - 2gA = (42n + 9) - (42n + 8) = 1 \ .$$

The first expression is divisible by g, so g is a divisor of 1. Hence $g = 1$, and the fraction is not reducible.

Second solution. The same result may be obtained by using Euclid's algorithm† for determining the greatest common divisor of the two numbers $21n + 4$ and $14n + 3$ as follows:

$$\begin{aligned} 21n + 4 &= 1 \cdot (14n + 3) + (7n + 1) \\ 14n + 3 &= 2 \cdot (7n + 1) + 1 \\ 7n + 1 &= 1 \cdot (7n + 1) + 0 \ ; \end{aligned}$$

the greatest common divisor is 1. It follows that the given fraction is irreducible for all natural numbers n.

1959/2. Since $\sqrt{2x - 1}$ is assumed to be real, we must have $x \geqslant 1/2$. Let

$$\sqrt{x + \sqrt{2x - 1}} + \sqrt{x - \sqrt{2x - 1}} = A \ .$$

We square both sides and obtain

$$(1) \qquad 2x + 2\sqrt{x^2 - (2x - 1)} = A^2 \ .$$

†For more on Euclid's algorithm, see for example *Continued Fractions*, by C. D. Olds, vol 9 of this NML series, 1963, pp. 16–17.

Since $x^2 - (2x - 1) = (x - 1)^2$, we can write (1) in the equivalent form

$$(1)' \qquad x + |x - 1| = A^2/2 .$$

(a) Suppose $A = \sqrt{2}$; then $A^2/2 = 1$, and we look for solutions $x \geqslant 1/2$ of

$$(2) \qquad x + |x - 1| = 1 .$$

For $1/2 \leqslant x \leqslant 1$, $x - 1 \leqslant 0$, so $|x - 1| = 1 - x$, and equation (2) becomes the identity

$$x + 1 - x = 1 \qquad \text{for } 1/2 \leqslant x \leqslant 1 .$$

If $1 < x$, then $x - 1 > 0$, so $|x - 1| = x - 1$, and (2) becomes

$$x + x - 1 = 2x - 1 = 1 ;$$

this equation is satisfied only if $x = 1$, which contradicts the assumption that $x > 1$. We conclude that $A = \sqrt{2}$ if and only if $1/2 \leqslant x \leqslant 1$.

In view of this result, we can confine attention to the range $x > 1$ in solving parts (b) and (c). Then $x + |x - 1| = 2x - 1$.

(b) If $A = 1$, then $2x - 1 = A^2/2 = 1/2$, so $x = 3/4$. Since this is not in the range $x > 1$, there are no solutions for $A = 1$.

(c) If $A = 2$, then $2x - 1 = A^2/2 = 2$, so $x = 3/2$.

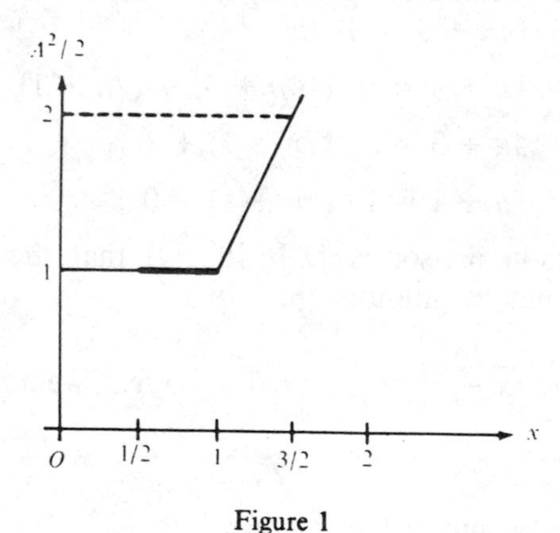

Figure 1

Remark. It is helpful to consider expression (1)′ for $A^2/2$ as the sum of the functions x and $|x - 1|$, to note that for $1/2 \leqslant x \leqslant 1$ this sum has the constant value 1, and for $x > 1$ it is the linear function $2x - 1$; see Figure 1. Cases (a), (b) and (c) correspond to function values 1, 1/2 and 2, and the solutions can be read off from the graph.

1959/3 First solution. In the given equation $a\cos^2 x + b\cos x + c = 0$, we substitute

$$\cos x = \pm\sqrt{\frac{1+\cos 2x}{2}}\,,$$

obtaining

$$a\left(\frac{1+\cos 2x}{2}\right) \pm b\sqrt{\frac{1+\cos 2x}{2}} + c = 0\,,$$

which simplifies to

$$(1)\quad a^2\cos^2 2x + (2a^2 + 4ac - 2b^2)\cos 2x + a^2 + 4ac + 4c^2 - 2b^2 = 0\,.$$

For $a = 4$, $b = 2$, $c = -1$, the given equation is

$$4\cos^2 x + 2\cos x - 1 = 0\,,$$

and the left member of (1) is $4(4\cos^2 2x + 2\cos 2x - 1)$, so that the quadratic equation for $\cos 2x$ has coefficients proportional to those of the given equation.

It follows that these quadratic equations have the same roots, i.e., that the pairs $\{\cos 2x_1, \cos 2x_2\}$ and $\{\cos x_1, \cos x_2\}$ are identical. Hence either

(i) $\cos 2x_1 = \cos x_1$, $\quad \cos 2x_2 = \cos x_2$ $\quad$ or

(ii) $\cos 2x_1 = \cos x_2$, $\quad \cos 2x_2 = \cos x_1$.

In the first case, $2x_1 = 360°k \pm x_1$, so that x_1 is a multiple of 120° and its cosine does not satisfy the equation. In the second case $2x_1 = 360k \pm x_2$, $2x_2 = 360°h \pm x_1$. When we use like signs in the last terms and solve this system, we are back in case (i); but if we use opposite signs and solve this system, we find that $5x_2 = 360°j$ and that the angles 72°, 144°, 216° and 288° satisfy our equation. Since the real roots of quadratic equations with integer coefficients can be constructed with straight edge and compasses, this analysis shows that a regular pentagon can be constructed with straight edge and compasses.

Second solution (due to Gerhard Arenstorf). Let $r_1 = \cos x_1$, $r_2 = \cos x_2$ be the roots of the quadratic equation $q(y) = ay^2 + by + c = 0$. Since $\cos 2x_1 = 2\cos^2 x_1 - 1$, $\cos 2x_2 = 2\cos^2 x_2 - 1$, the roots R_1 and R_2 of the desired quadratic equation

$$Q(z) = Az^2 + Bz + C = 0$$

are

$$R_1 = 2r_1^2 - 1 \qquad \text{and} \qquad R_2 = 2r_2^2 - 1\,.$$

Their sum is

$$R_1 + R_2 = 2(r_1^2 + r_2^2) - 2 = 2\left[(r_1 + r_2)^2 - 2r_1r_2\right] - 2 = -\frac{B}{A}\,,$$

and since $r_1 + r_2 = -b/a$, $r_1 r_2 = c/a$, we have

$$R_1 + R_2 = 2\left[\left(\frac{b}{a}\right)^2 - 2\frac{c}{a}\right] - 2 = 2\left(\frac{b^2}{a^2}\right) - 4\frac{c}{a} - 2 = -\frac{B}{A} .$$

The product of the roots of $Q(z) = 0$ is

$$R_1 R_2 = (2r_1^2 - 1)(2r_2^2 - 1) = 4r_1^2 r_2^2 - 2(r_1^2 + r_2^2) + 1$$

$$= 4\frac{c^2}{a^2} - 2\left[\left(\frac{b}{a}\right)^2 - 2\frac{c}{a}\right] + 1 = \frac{C}{A} .$$

Setting $A = a^2$, we conclude that

$$Q(z) = az^2 + (2a^2 + 4ac - 2b^2)z + 4c^2 - 2b^2 + 4ac + a^2 = 0 .$$

1959/4 First solution. Let the given hypotenuse be the diameter AB of the semicircle in Figure 2. Every triangle ACB inscribed in this semi-circle has a right angle at C and a median CO of length $c/2$. Our task is to locate the point C in such a way that CO is the geometric mean of $a = CB$ and $b = AC$, i.e.

$$\text{(1)} \qquad \frac{a}{c/2} = \frac{c/2}{b} \quad \text{or} \quad 4ab = c^2 .$$

We accomplish this by determining the altitude h from the vertex C of the desired triangle. On the one hand, the area of this triangle is $ab/2$ and also $ch/2$; hence $ab = ch$. On the other hand, equation (1) must hold. Thus

$$ch = ab = c^2/4 , \qquad \text{whence } h = c/4 .$$

We construct a line parallel to the diameter AB and at a distance $c/4$ above it. This line intersects the semicircle in two points C and D, which are the vertices of right triangles satisfying the requirements of the problem.

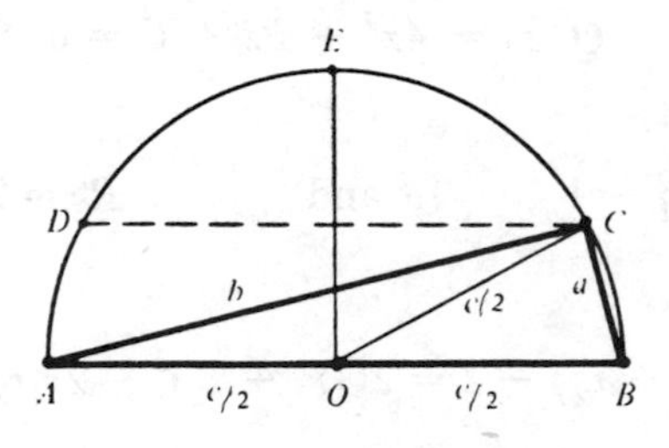

Figure 2

Second solution. By the law of cosines, applied to $\triangle COB$ with $\angle COB = \theta$,

$$a^2 = \frac{c^2}{4} + \frac{c^2}{4} - 2\frac{c^2}{4}\cos\theta = c^2\left(\frac{1-\cos\theta}{2}\right),$$

and applied to $\triangle AOC$, where $\angle AOC = \pi - \theta$,

$$b^2 = \frac{c^2}{4} + \frac{c^2}{4} + 2\frac{c^2}{4}\cos\theta = c^2\left(\frac{1+\cos\theta}{2}\right).$$

Hence

$$a^2b^2 = c^4\frac{1-\cos^2\theta}{4} = \frac{c^4}{4}\sin^2\theta, \qquad \text{so} \quad ab = \frac{c^2}{2}\sin\theta\ .$$

But the conditions of the problem imply that $ab = c^2/4$. Therefore $\sin\theta = 1/2$, $\theta = \pi/6$ or $\theta = 5\pi/6$.

1959/5. (a) Consider Figure 3. Draw segments AN, NF, BC and CN. Then $\angle ANM = 45°$ and $\angle MNF = 135°$. Therefore $\angle ANF = \angle ANM + \angle MNF = 45° + 135° = 180°$, so N lies on AF. Since AC is a diameter, $\angle ANC = 90°$; $\angle ANB = \angle ANM + \angle MNB = 45° + 45° = 90°$. Hence C lies on BN, and thus AF and BC intersect at N. Therefore $N = N'$. This proves part (a) of the problem.

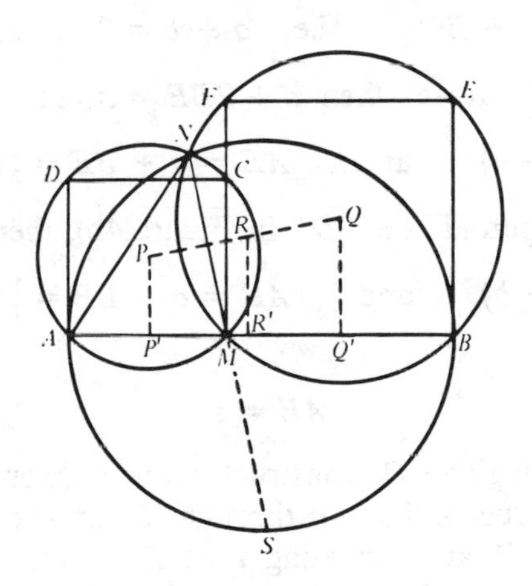

Figure 3

(b) Line NM bisects $\angle ANB$ since $\angle ANM = \angle MNB = 45°$. Therefore NM, when extended, bisects the lower semicircle with diameter AB. Denote the midpoint of this semicircle by S. Since $AN \perp BN$ for every M on AB, N traces out the upper semicircle with diameter AB as M moves from A to B, while S is always the midpoint of the lower semicircle. This shows that S does not depend on the choice of M, and so proves part (b) of the problem.

(c) Let PP', QQ', RR' be perpendiculars to AB from P, Q, and the midpoint R of segment PQ, respectively. Since RR' is the median of trapezoid $P'Q'QP$,

$$RR' = \frac{1}{2}(PP' + QQ') = \frac{1}{2}\,\frac{AM + MB}{2} = \frac{AB}{4}\ .$$

This shows that the distance from R to AB is constant. When $M = A$, we have $P' = A$ and $AQ' = AB/2$, so $AR' = AB/4$. When $M = B$, we have $P'B = AB/2$ and $Q' = B$, so $R'B = AB/4$. Thus the locus of R is a line segment of length $AB/2$, parallel to AB and at distance $AB/4$ above it; it is bisected by the perpendicular bisector of AB.

1959/6. We first list some facts which are basic to the construction. The parallel sides AB and CD of the desired trapezoid $ABCD$ must both be parallel to p.† The distance CE between them (see Figure 4) is equal to the diameter of the inscribed circle. The distance AC between the given points A and C is known. Denote the still unknown distances AB and CD by a and b, respectively, and the equal distances AD and BC by s. Since the two tangents from a vertex of the quadrilateral to the inscribed circle are equal,

$$AB + CD = AD + BC\ , \qquad \text{i.e.}\quad a + b = 2s\ , \quad \text{so}\quad s = \tfrac{1}{2}(a + b)\ .$$

If $a < b$ (as in Figure 4a), then $a + 2BE = b$, so

$$BE = \tfrac{1}{2}(b - a) \qquad \text{and} \qquad AE = a + BE = \tfrac{1}{2}(a + b) = s\ .$$

If $a > b$ (see trapezoid $AB'CD'$ in Figure 4b), then $a - 2BE = b$, so

$$BE = \tfrac{1}{2}(a - b) \qquad \text{and} \qquad AE = a - BE = \tfrac{1}{2}(a + b) = s\ .$$

Thus in both cases

$$AE = s\ .$$

Construction: In plane P, construct a line l through A parallel to p; in plane Q, construct a line m through C parallel to p. Construct a perpendicular to CD at C meeting l at E. With AE as radius and C as center, draw an arc of a circle intersecting line AE at B. With the same radius and A as center, draw an arc of a circle intersecting m at D. Both desired vertices B and D are now located.

If $CE < AE$, two solutions exist ($ABCD$ and $AB'CD'$ in Figures 4). If $CE = AE$, there is exactly one solution, a square. If $CE > AE$, the problem has no solution.

†For if M is a point of p, there is a unique line l through M parallel to AB and CD. Since l lies in both the planes P and Q, it follows that $l = p$.

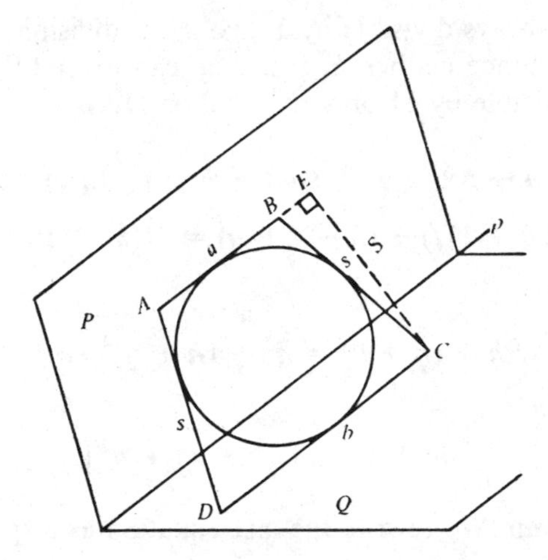

Figure 4a

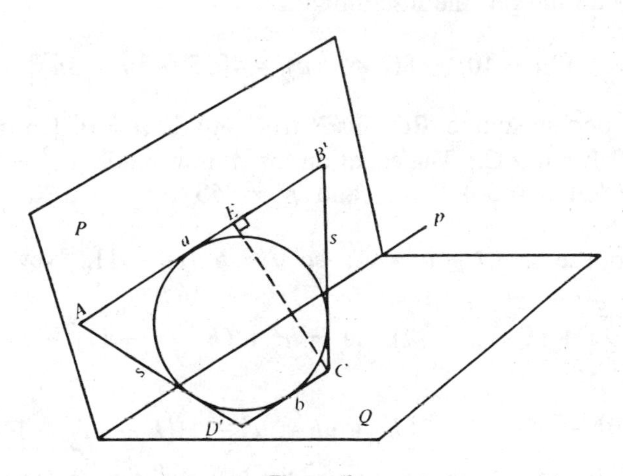

Figure 4b

Second International Olympiad, 1960

1960/1. Let

$$N = 100h + 10t + u \ ;$$

the digits h, t, u are to be determined so that

$$N = 11(h^2 + t^2 + u^2) \ .$$

We note first that N can be written as a sum of two terms,

$$N = (99h + 11t) + (h - t + u) \ ;$$

the first term is always divisible by 11, so N is divisible by 11 if and only if $h - t + u$ is. Since neither h, t nor u can exceed 9, the combination $h - t + u$ is divisible by 11 only if it is 0 or 11.

(a) Suppose $h - t + u = 0$, so $t = h + u$. In this case the relation

$$(99h + 11t) + (h - t + u) = 11(h^2 + t^2 + u^2) \tag{1}$$

becomes

$$9h + (h + u) = h^2 + (h + u)^2 + u^2 ,$$

or

$$10h + u = 2[h^2 + uh + u^2] ,$$

so u must be even. We rewrite the last equation as a quadratic in h:

$$2h^2 + (2u - 10)h + 2u^2 - u = 0 .$$

Since h is an integer, the discriminant

$$(2u - 10)^2 - 8(2u^2 - u) = 4[25 - 8u - 3u^2]$$

must be a perfect square. But this is true only if $u = 0$. [In fact $25 - 8u - 3u^2 < 0$ for $u \geqslant 2$.] The equation for h now reads $2h^2 - 10h = 0$, so $h = 5$, $t = h + u = 5 + 0 = 5$, and $N = 550$.

(b) Suppose $h - t + u = 11$; so $t = h + u - 11$. Now relation (1) yields

$$9h + (h + u - 11) + 1 = h^2 + (h + u - 11)^2 + u^2 ,$$

or

$$10h + u - 10 = 2[h^2 + uh + u^2 - 11(h + u)] + 121 ,$$

so u must be odd. Again we rewrite the last equation in the form

$$2h^2 + (2u - 32)h + 2u^2 - 23u + 131 = 0$$

and require that its discriminant,

$$(2u - 32)^2 - 8(2u^2 - 23u + 131) = 4[-3u^2 + 14u - 6]$$

be a perfect square. Since u is odd, we test 1 (which does not yield a perfect square), 3, which yields the square 36, and find that all larger odd u yield a negative discriminant. If $u = 3$, our quadratic in h becomes

$$2h^2 - 26h + 80 = 0 \qquad \text{or} \qquad h^2 - 13h + 40 = (h - 5)(h - 8) = 0 .$$

When $h = 5$, $t = 5 + 3 - 11 = -3$, not admissible. When $h = 8$, $t = 8 + 3 - 11 = 0$, so $N = 803$.

The only two numbers satisfying the conditions of the problem are therefore 550 and 803.

1960/2. The left side of the given inequality is defined if $x \neq 0$ and real if $x \geqslant -1/2$. Assume that x satisfies these conditions. We multiply numerator and denominator of the left member by $(1+\sqrt{1+2x})^2$ and obtain the equivalent inequality

$$(1+\sqrt{1+2x})^2 < 2x+9 .$$

Putting $y=\sqrt{1+2x}$, we have $(1+y)^2 < y^2+8$, which simplifies to $y<7/2$. Hence $1+2x<49/4$, and $x<45/8$. The complete solution of the problem therefore consists of all x satisfying

$$-\frac{1}{2} \leqslant x < \frac{45}{8}, \qquad x \neq 0 .$$

1960/3. First solution (due to A. Zisook). Let P and Q denote the endpoints of the segment containing the midpoint O of BC. Let H be the foot of the altitude from A. In Figure 5, denote $\angle HAQ$ by β, $\angle HAP$ by γ and BH by s. Then $\alpha=\beta-\gamma$, and

$$\tan\alpha = \tan(\beta-\gamma) = \frac{\tan\beta - \tan\gamma}{1+\tan\beta\tan\gamma} .$$

Now

$$\tan\beta = \frac{HQ}{AH} = \frac{BO-BH+OQ}{h} = \frac{a-2s}{2h} + \frac{a}{2nh}$$

$$\tan\gamma = \frac{HP}{AH} = \frac{BO-BH-OP}{h} = \frac{a-2s}{2h} - \frac{a}{2nh},$$

so

$$\tan\alpha = \frac{a/nh}{1+\left[a^2(n^2-1)-4n^2s(a-s)\right]/4n^2h^2} . \tag{2}$$

Since h is the altitude of a right triangle,

$$\frac{s}{h} = \frac{h}{a-s}, \qquad \text{so} \qquad s(a-s)=h^2 .$$

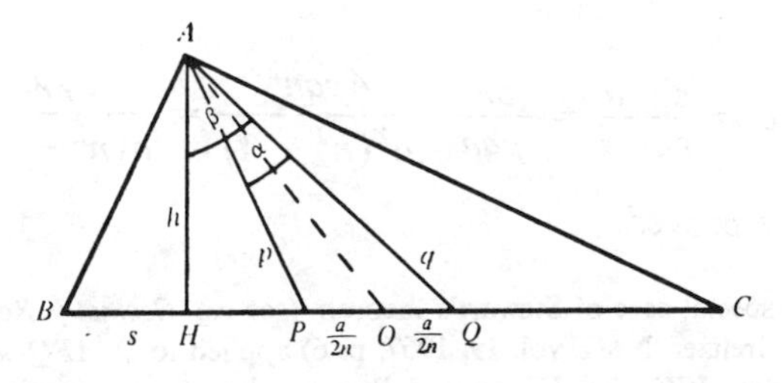

Figure 5

We substitute the last expression into (2), simplify, and obtain

$$\tan\alpha = \frac{4nh}{a(n^2-1)} .$$

Second solution. Denote the lengths of AP and AQ by p and q respectively; see Figure 5. Thus

$$AH = h \qquad AO = \frac{a}{2}, \qquad OP = OQ = \frac{a}{2n} .$$

We now apply the law of cosines to $\triangle AOP$ and $\triangle AOQ$ to express p^2 and q^2 in terms of the given quantities:

$$p^2 = \frac{a^2}{4} + \frac{a^2}{4n^2} - 2\frac{a}{2}\frac{a}{2n}\cos\angle AOP ,$$

$$q^2 = \frac{a^2}{4} + \frac{a^2}{4n^2} - 2\frac{a}{2}\frac{a}{2n}\cos\angle AOQ .$$

Since $\angle AOP$ and $\angle AOQ$ are supplementary angles, we have $\cos\angle AOP = -\cos\angle AOQ$. Therefore, adding the above equations, we obtain

$$(1) \qquad p^2 + q^2 = \frac{a^2(n^2+1)}{2n^2} .\dagger$$

The area of $\triangle APQ$ may be expressed as $\frac{1}{2}PQ\cdot h = ah/2n$, and also as $\frac{1}{2}pq\sin\alpha$. Equating these and solving for $\sin\alpha$, we get

$$\sin\alpha = \frac{ah}{pqn} .$$

By the law of cosines, $PQ^2 = (a/n)^2 = p^2 + q^2 - 2pq\cos\alpha$, so

$$\cos\alpha = \frac{1}{2pq}\left[p^2 + q^2 - \frac{a^2}{n^2}\right] .$$

Substituting from (1) for p^2+q^2, we obtain

$$\cos\alpha = \frac{a^2(n^2-1)}{4pqn^2} ,$$

and

$$\tan\alpha = \frac{\sin\alpha}{\cos\alpha} = \frac{ah}{pqn}\cdot\frac{4pqn^2}{a^2(n^2-1)} = \frac{4nh}{a(n^2-1)} ,$$

which was to be proved.

†This result is a special case of Stewart's theorem (see e.g. *Geometry Revisited* by H. S. M. Coxeter and S. L. Greitzer, NML vol. 19, 1967, p. 6) applied to $\triangle APQ$ with cevian AO: In $\triangle UVW$ with sides $VW = u$, $UV = w$, $UW = v$, let X divide side VW into parts $VX = d$, $XW = e$, and let $UX = f$. Then $u(f^2 + de) = v^2d + w^2e$.

1960/4. Construct a right triangle HAM with m_a as hypotenuse and h_a as one leg (for example by using $AM = m_a$ as diameter of a circle G and locating the point H by drawing a circular arc with center at A and radius h_a, see Figure 6). Now if M is the midpoint of side BC of the desired triangle and BK the given altitude h_b, then MT, perpendicular to side AC, is parallel to BK and half as long. Hence T is an intersection point of the circle G and a circular arc centered at M with radius $\frac{1}{2}h_b$†. Now extend HM and AT; their intersection is C; find B by laying off a distance $MB = MC$ along the line MH.

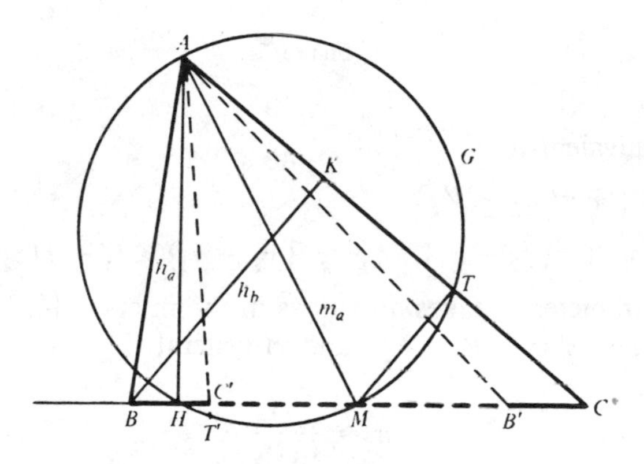

Figure 6

1960/5. Suppose for simplicity that the cube is a unit cube. Place it on an x, y, z coordinate system with D at the origin, and three of its edges along the coordinate axes, so that its vertices are at

$$D = (0,0,0)\,,\quad A = (1,0,0)\,,\quad B = (1,1,0)\,,\quad C = (0,1,0)\,,$$
$$D' = (0,0,1)\,,\quad A' = (1,0,1)\,,\quad B' = (1,1,1)\,,\quad C' = (0,1,1)\,;$$

see Figure 7a.

We shall make repeated use of the following facts: If $V_1 = (x_1, y_1, z_1)$ and $V_2 = (x_2, y_2, z_2)$ are any two points in space, then all points V on the line segment connecting them can be represented by

$$\begin{aligned} V &= (1-\tau)V_1 + \tau V_2 = (1-\tau)(x_1, y_1, z_1) + \tau(x_2, y_2, z_2) \\ &= ((1-\tau)x_1 + \tau x_2,\ (1-\tau)y_1 + \tau y_2,\ (1-\tau)z_1 + \tau z_2) \end{aligned}$$

† There are two such intersections: one on the semicircular arc AM not containing H and another, T', on the semicircle containing H (see Figure 6). Using T' instead of T leads to a second solution $AB'C'$ of the problem.

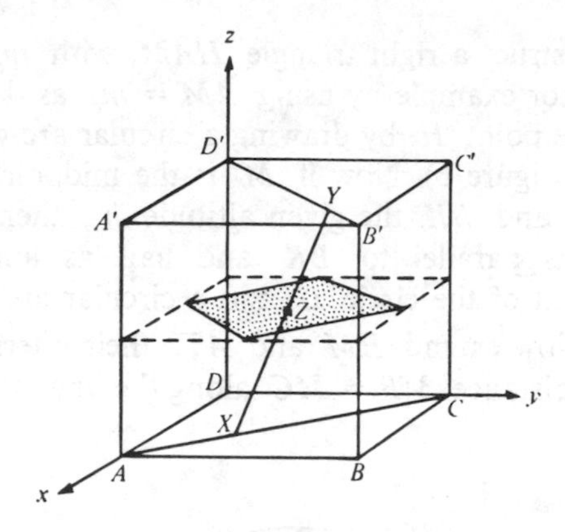

Figure 7a

which is equivalent to

$$\begin{aligned} V &= V_1 + \tau(V_2 - V_1) \\ &= (x_1 + \tau(x_2 - x_1),\ \ y_1 + \tau(y_2 - y_1),\ \ z_1 + \tau(z_2 - z_1)) \ . \end{aligned}$$

Here the parameter τ takes all values in the interval $[0, 1]$. When $\tau = 0$, $V = V_1$; when $\tau = 1$, $V = V_2$; and in general

$$\tau = \frac{V_1V}{V_1V_2} \ .$$

To solve our problem, we first recall that X lies on AC and Y on $D'B'$. By the above,

$$X = (1 - s)A + sC = (1 - s, 0, 0) + (0, s, 0) = (1 - s, s, 0) \ , 0 \leqslant s \leqslant 1$$

$$Y = (1 - t)D' + tB' = (0, 0, 1 - t) + (t, t, t) = (t, t, 1) \ , \quad 0 \leqslant t \leqslant 1 \ .$$

(a) The midpoint of XY is $P = (1 - \tau)X + \tau Y$ with $\tau = 1/2$, so its coordinates are

$$P = \left(\frac{1 - s}{2}, \frac{s}{2}, 0\right) + \left(\frac{t}{2}, \frac{t}{2}, \frac{1}{2}\right) = \left(\frac{1 - s + t}{2},\ \ \frac{s + t}{2},\ \ \frac{1}{2}\right).$$

The z-coordinate of P is always $1/2$, so P lies in the plane $z = 1/2$ parallel to and halfway between the planes $z = 0$ and $z = 1$. The sum and difference of its x and y coordinates are

$$x + y = \frac{1 + 2t}{2} = t + \frac{1}{2} \qquad \text{and} \qquad x - y = \frac{1 - 2s}{2} = \frac{1}{2} - s \ ,$$

respectively. Since $0 \leqslant t \leqslant 1$ and $0 \leqslant s \leqslant 1$, we have

$$\frac{1}{2} \leqslant x + y \leqslant \frac{3}{2} \qquad \text{and} \qquad -\frac{1}{2} \leqslant x - y \leqslant \frac{1}{2} \ .$$

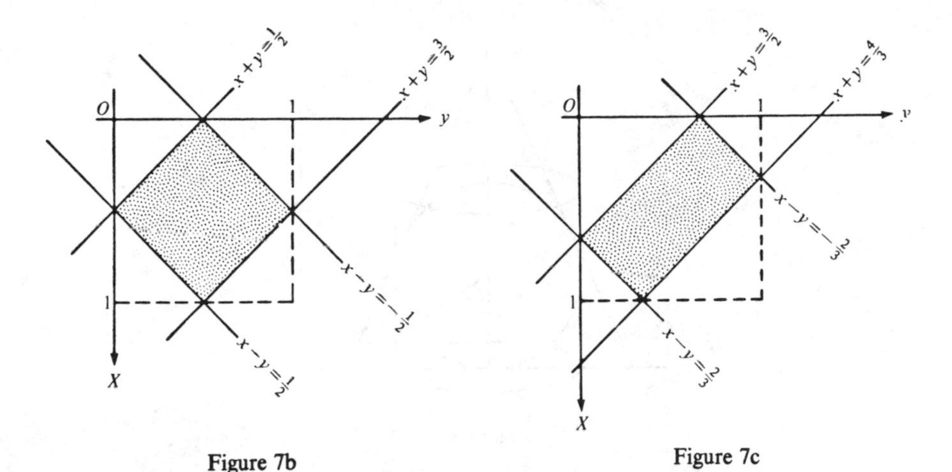

Figure 7b

Figure 7c

Figure 7b shows the strips in the plane $z = 1/2$ where each of the above inequalities holds; the shaded square where both inequalities hold is the locus of P. The square has side $\sqrt{2}/2$.

(b) The coordinates of the point Z such that $ZY/XZ = 2$ or $XZ/XY = 1/3$ are $(1-\tau)X + \tau Y$ with $\tau = 1/3$; hence

$$Z = \frac{2}{3}(1-s, s, 0) + \frac{1}{3}(t, t, 1) = \left(\frac{2-2s+t}{3}, \quad \frac{2s+t}{3}, \quad \frac{1}{3}\right).$$

Thus Z lies on the plane $z = 1/3$. The sum and difference of its x and y-coordinates are

$$x + y = \frac{2+2t}{3} \qquad \text{and} \qquad x - y = \frac{2-4s}{3},$$

so

$$\frac{2}{3} \leqslant x + y \leqslant \frac{4}{3} \qquad \text{and} \qquad -\frac{2}{3} \leqslant x - y \leqslant \frac{2}{3}.$$

The strips in the plane $z = 1/3$ where these inequalities hold are pictured in Figure 7c; their intersection (shaded rectangle) is the locus of Z.

1960/6. Solution (due to C. Hornig). Denote the axis of the cone by VM, the radius of its base by r, the radius of the inscribed sphere by s and its center by O; see Figure 8. Since AM and AV are two tangents to the sphere from the same point A, we have $\angle MAO = \angle OAV$; denote these two angles by θ. Then the radius of the base of the cylinder is $s = r \tan\theta$, and the altitude of the cone is $h = r \tan 2\theta$.

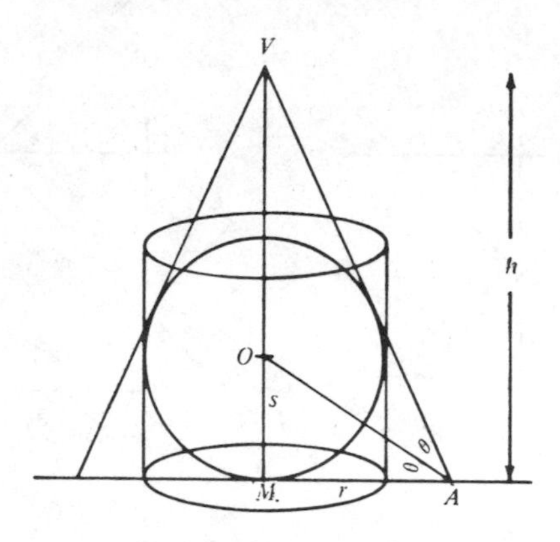

Figure 8

Now the volume of the cone is

$$V_1 = \frac{1}{3}\pi r^2 h = \frac{1}{3}\pi r^3 \tan 2\theta = \frac{2\pi r^3 \tan\theta}{3(1-\tan^2\theta)},$$

and that of the cylinder is

$$V_2 = 2\pi s^3 = 2\pi r^3 \tan^3\theta .$$

Their ratio is

$$(1) \qquad \frac{V_1}{V_2} = \frac{1}{3\tan^2\theta(1-\tan^2\theta)} = k .$$

(a) To prove that $V_1 \neq V_2$, we shall show that $k \neq 1$. We rewrite (1) as the quadratic (in $\tan^2\theta$)

$$\tan^4\theta - \tan^2\theta + \frac{1}{3k} = 0 ,$$

and observe that its discriminant, $1 - 4/3k$, is nonnegative only if $k \geqslant 4/3$. This shows that $k \neq 1$ and thus proves part (a).

Alternatively, the arithmetic mean-geometric mean inequality† $\sqrt{ab} \leqslant (a+b)/2$ can be applied with $a = \tan^2\theta$, $b = 1 - \tan^2\theta$ to show that $\tan^2\theta(1-\tan^2\theta) \leqslant 1/4$, and hence $k \geqslant 4/3$. Equality holds if and only if $a = b$, i.e. $\tan^2\theta = 1/2$.

(b) The smallest value of k for which $\tan^2\theta$ is real is $4/3$, and when $k = 4/3$, $\tan^2\theta = 1/2$, so $\tan\theta = 1/\sqrt{2}$. Hence

$$\tan 2\theta = \frac{2\tan\theta}{1-\tan^2\theta} = 2\sqrt{2} = \frac{h}{r} .$$

Construct a square of side r; then h is twice its diagonal. Right triangle VMA can now be constructed with sides h and r. Finally, construct the desired angle at V by doubling $\angle MVA$.

†See p. 196

1960/7. In Figure 9, let MN be the axis of symmetry of the trapezoid $ABCD$, and P a point on MN such that $\angle BPC$ is a right angle. Right triangles CPM and PBN are similar, since corresponding sides are perpendicular; hence $MP/NB = MC/NP$. Using the abbreviations $MP = d$, $CD = a$, $AB = c$, and $MN = h$, we may write this in the form $d(h-d) = (a/2)(c/2)$, so that $4d^2 - 4hd + ac = 0$. Therefore

$$d = \frac{h}{2} \pm \frac{1}{2}\sqrt{h^2 - ac} \ .$$

Since the sum of the roots of the equation is h, one of the roots is equal to PM, the other to PN.

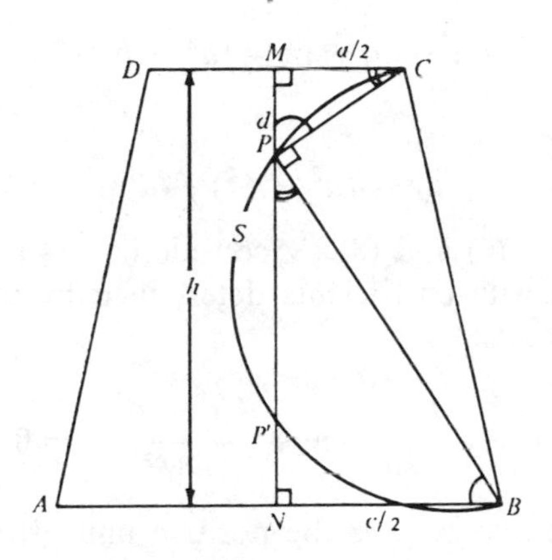

Figure 9

(1) If $h^2 - ac < 0$, there is no point P;

(2) if $h^2 - ac = 0$, there is one point P, the midpoint of altitude MN;

(3) if $h^2 - ac > 0$, there are two points P and P' which satisfy the conditions. Clearly $NP' = PM$.

Remark. The geometric interpretation of these three cases is as follows. In case (1), the circle S with diameter BC does not intersect MN, in case (2) S is tangent to MN, while in case (3) S meets MN in two points P and P'.

Third International Olympiad, 1961

1961/1 First solution. The given equations are

(1) $$x + y + z = a$$

(2) $$x^2 + y^2 + z^2 = b^2$$

(3) $$xy = z^2 \ .$$

In (2) we substitute xy for z^2 from (3), obtaining

(4) $$x^2 + xy + y^2 = b^2 .$$

We rewrite (1) in the form $(x + y) - a = -z$, square it, substitute xy for z^2 from (3) and obtain

(5) $$x^2 + xy + y^2 - 2a(x + y) = -a^2 .$$

We subtract (5) from (4) and find that

(6) $$x + y = (a^2 + b^2)/2a ;$$

hence

(7) $$z = a - (x + y) = (a^2 - b^2)/2a .$$

By (3) and (7),

(8) $$xy = (a^2 - b^2)^2/4a^2 .$$

We solve the system (6) and (8) by considering x and y as roots of a quadratic equation with coefficients determined by the sum (6) and the product (8) of the roots:

(9) $$w^2 - \frac{a^2 + b^2}{2a} w + \frac{(a^2 - b^2)^2}{4a^2} = 0 .$$

The requirement that x, y, z be positive implies, by (1), that $a > 0$, and by (7), that $a^2 > b^2$. Moreover, the requirement that x and y be real and distinct forces the discriminant of (9),

$$\triangle = \frac{1}{4a^2}(3a^2 - b^2)(3b^2 - a^2) ,$$

to be positive. Since $3a^2 - b^2 > 0$, it follows that $3b^2 - a^2 > 0$, so that $3b^2 > a^2$. All these inequalities can be summarized as

$$|b| < a < \sqrt{3}\,|b| ,$$

and if they hold, the roots of (9),

$$x = \frac{a^2 + b^2}{4a} \pm \frac{\sqrt{\triangle}}{2} , \qquad y = \frac{a^2 + b^2}{4a} \mp \frac{\sqrt{\triangle}}{2}$$

together with their geometric mean, $z = \sqrt{xy} = \dfrac{a^2 - b^2}{2a}$, yield positive, distinct solutions of the problem. (z is distinct from x and y since the geometric mean of two distinct numbers lies between them. See p. 200.)

Our knowledge of the relations between the roots and the coefficients of a quadratic equation led us to use the sum (6) and the product (8) of x and y to write the quadratic equation (9) satisfied by x and y. The symmetry in x, y, z of the given equations (1) and (2) encourages us to try solving the problem by regarding x, y, z as roots of a cubic, as follows:

Second solution. Let x, y, z be the roots of

$$p^3 + s_1 p^2 + s_2 p + s_3 = 0 \; ; \tag{10}$$

then

$$x + y + z = -s_1 , \qquad xy + yz + xz = s_2 , \qquad xyz = -s_3 .$$

Using (1), (2) and (3), we obtain

$$s_1 = -a , \qquad (x + y + z)^2 = a^2 = b^2 + 2s_2 , \qquad xyz = z^3 = -s_3 ,$$

respectively, so that

$$s_1 = -a , \qquad s_2 = \tfrac{1}{2}(a^2 - b^2) , \qquad s_3 = -z^3 .$$

The cubic (10) may therefore be written as

$$p^3 - ap^2 + \frac{1}{2}(a^2 - b^2)p - z^3 = 0 . \tag{11}$$

Since z is a root, we find on substituting $p = z$ into (11), that

$$-az^2 + \frac{1}{2}(a^2 - b^2)z = -z\left[az - \frac{1}{2}(a^2 - b^2)\right] = 0 ,$$

so $z = 0$ or $z = (a^2 - b^2)/2a$. If $z = 0$, then the corresponding solution x, y, z would *not* consist of positive numbers, as required. But if $z = (a^2 - b^2)/2a$, then by (1) $x + y = a - z = (a^2 + b^2)/2a$, and by (3), $xy = (a^2 - b^2)^2/4a^2$. We solve these for x and y by the method given in the first solution.

Note. The necessity of the conditions

$$|b| < a < \sqrt{3}\,|b|$$

for x, y, z positive and distinct may be seen in these two ways:

(i) The distance from the plane $x + y + z = a$ to the origin is $a/\sqrt{3}$, and the radius of the sphere $x^2 + y^2 + z^2 = b^2$ about the origin is $|b|$. These surfaces have points in common if and only if $|b| \geqslant a/\sqrt{3}$. If equality holds, the sphere is tangent to the plane at a point with equal coordinates, so $0 < a < \sqrt{3}\,|b|$ is necessary for x, y, z distinct and positive.

(ii) To the two positive numbers x and y, apply the arithmetic-geometric mean inequality†

$$\frac{1}{2}(x + y) \geqslant \sqrt{xy} ,$$

†See p. 196

using relations (6) and (8) above. The result is $(a^2 + b^2)/4a > a^2b^2/2a$, where the inequality is strict since $x \neq y$. This simplifies to $a^2 < 3b^2$, or since a is positive, to $0 < a < \sqrt{3}\,|b|$.

1961/2 First solution (due to G. Arenstorf). Of the three altitudes of a triangle, at least one lies inside the triangle; denote its length by h, and let it divide the side to which it is drawn into segments m and n (see Figure 10a). Then the squares of the other two sides are $h^2 + m^2$ and $h^2 + n^2$, and the area T is

$$T = \frac{1}{2}(m + n)h \ .$$

The inequality to be proved takes the form

$$2h^2 + m^2 + n^2 + (m + n)^2 \geqslant 2\sqrt{3}\,(m + n)h \ ,$$

which is equivalent to the inequality

$$(1) \qquad h^2 - \sqrt{3}\,(m + n)h + n^2 + m^2 + mn \geqslant 0 \ .$$

The left member is a quadratic in h; we shall call it $Q(h)$. By completing the square we can bring it into the form

$$Q(h) = \left[h - \frac{\sqrt{3}}{2}(m + n)\right]^2 + \left[\frac{1}{2}(m - n)\right]^2 ,$$

the sum of two squares. Therefore $Q(h)$ is never negative, and inequality (1) is satisfied for all h. $Q(h) = 0$ if and only if $m = n$ and $h = \sqrt{3}\,m$. In this case the altitude from A bisects the base BC and has length $(\sqrt{3}\,/2)BC$. Hence $\triangle ABC$ is equilateral when $a^2 + b^2 + c^2 = 4\sqrt{3}\,T$.

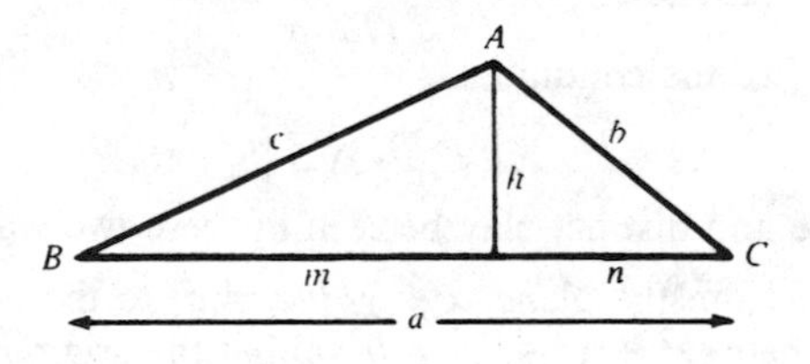

Figure 10a

Second solution. Denote the perimeter of the triangle by p:

$$p = a + b + c \ .$$

According to the isoperimetric theorem for triangles†, among all triangles

†For a proof, see e.g. *Geometric Inequalities* by N. D. Kazarinoff, NML vol. 4, 1961, p. 38.

with a fixed perimeter, the equilateral triangle has the largest area. The area of an equilateral triangle with side $p/3$ is $(p/3)^2\sqrt{3}/4$, so

$$\text{(1)} \qquad T \leqslant \left(\frac{p}{3}\right)^2 \frac{\sqrt{3}}{4} .$$

Moreover, the sum of the identities

$$p^2 = (a+b+c)^2 = a^2 + b^2 + c^2 + 2ab + 2bc + 2ac$$

and

$$(a-b)^2 + (b-c)^2 + (c-a)^2 = 2a^2 + 2b^2 + 2c^2 - 2ab - 2bc - 2ac$$

is

$$p^2 + (a-b)^2 + (b-c)^2 + (c-a)^2 = 3(a^2 + b^2 + c^2) ,$$

whence

$$\text{(2)} \qquad p^2 \leqslant 3(a^2 + b^2 + c^2) ,$$

with equality if and only if $a = b = c$. From (1) and (2) it follows that

$$T \leqslant \frac{a^2 + b^2 + c^2}{3} \frac{\sqrt{3}}{4} ,$$

which is equivalent to $a^2 + b^2 + c^2 \geqslant 4\sqrt{3}\, T$. Since equality in (1) and (2) holds if and only if $a = b = c$, it holds in the final inequality if and only if the triangle is equilateral.

Third solution. (i) Suppose that in $\triangle ABC$, $\angle A \geqslant 120°$, and construct equilateral $\triangle PBC$ on side BC, as shown in Figure 10b. The

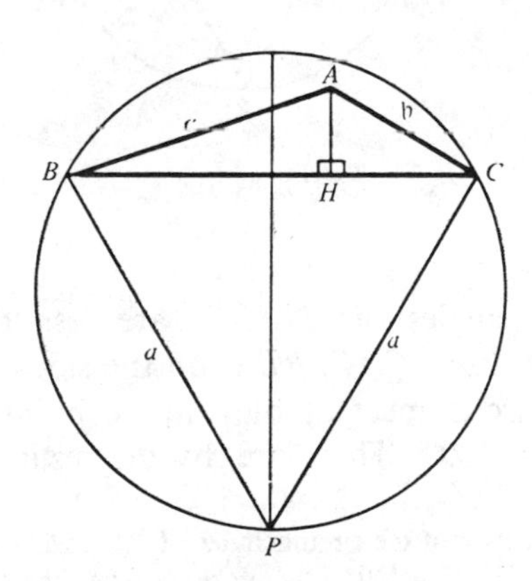

Figure 10b

diameter of the circumcircle of $\triangle PBC$ is $2a/\sqrt{3}$, and the altitude AH of $\triangle ABC$ satisfies $AH \leqslant a\sqrt{3}/6$. It follows that

$$\frac{(PBC)}{(ABC)} \geqslant \frac{a\sqrt{3}/2}{a\sqrt{3}/6} = 3,$$

where (XYZ) denotes area of $\triangle XYZ$, and so

$$(3) \qquad (PBC) = \frac{a^2\sqrt{3}}{4} \geqslant 3T,$$

where equality holds if and only if A is on the circumcircle of $\triangle PBC$, and $AB = AC$. We now strengthen inequality (3) by adding the positive terms $b^2\sqrt{3}/4$ and $c^2\sqrt{3}/4$ to the left member, obtaining

$$\frac{\sqrt{3}}{4}(a^2 + b^2 + c^2) > 3T, \qquad a^2 + b^2 + c^2 > 4\sqrt{3}\,T,$$

which was to be shown.

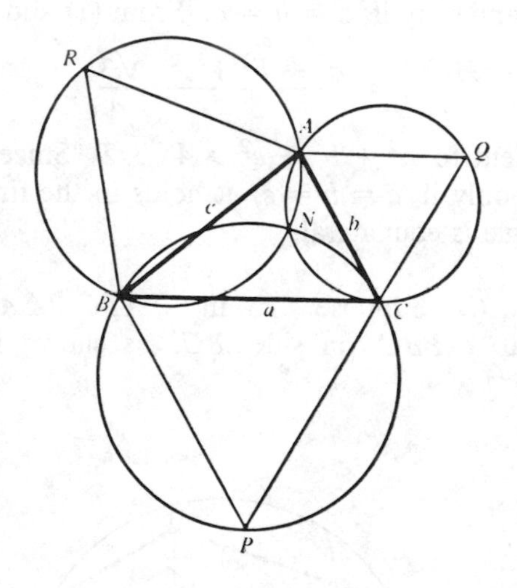

Figure 10c

(ii) Suppose all angles of $\triangle ABC$ are less than $120°$. Construct equilateral triangles PBC, QAC, RBA on the sides of $\triangle ABC$, see Figure 10c. Their circumcircles meet at the unique point† N, and $\angle ANB = \angle BNC = \angle CNA = 120°$. Therefore, by the result of (i) above

†Let N be the intersection of the circumcircles of $\triangle RAB$ and $\triangle QAC$. Then $\angle ANC = \angle ANB = 120°$, so $\angle BNC = 120°$; therefore N is on the circumcircle of $\triangle PBC$. [See also "Geometry Revisited" by H.S.M. Coxeter and S.L. Greitzer, NML vol. 19, 1967, p. 61.]

$$a^2\sqrt{3}\,/4 \geqslant 3(NBC)\ ,\quad b^2\sqrt{3}\,/4 \geqslant 3(NCA)\ ,\quad c^2\sqrt{3}\,/4 \geqslant 3(NAB)\ .$$

Adding, we find that

$$(a^2 + b^2 + c^2)\sqrt{3}\,/4 \geqslant 3T\ ,$$

so

$$a^2 + b^2 + c^2 \geqslant 4\sqrt{3}\,T\ .$$

Equality occurs when $NA = NB = NC$, i.e. when $\triangle ABC$ is equilateral.

1961/3. Since $\cos^n x - \sin^n x$ is periodic with period 2π, it suffices to find all the solutions in the interval $-\pi < x \leqslant \pi$. All other solutions are then obtained by adding integral multiples of 2π to these.

First we shall treat the case $n = 1$. We recall that any linear combination $A\cos x + B\sin x$ of $\cos x$ and $\sin x$ can be expressed as a pure sine, $C\sin(x + \theta)$. To find C and θ in terms of A and B, we use the addition formula

$$\begin{aligned} C\sin(x + \theta) &= C[\,\sin x\cos\theta + \cos x\sin\theta] \\ &= C\sin\theta\cos x + C\cos\theta\sin x\ . \end{aligned}$$

So $A = C\sin\theta$, $B = C\cos\theta$; $C^2 = A^2 + B^2$, $\tan\theta = B/A$. In particular, when $A = 1$, $B = -1$, we obtain

$$\cos x - \sin x = \sqrt{2}\,\sin\left(x + \frac{3\pi}{4}\right)\ .$$

Hence $\cos x - \sin x = 1$ if and only if

$$\sin\left(x + \frac{3\pi}{4}\right) = \frac{1}{\sqrt{2}}\ ,\quad x + \frac{3\pi}{4} = \frac{\pi}{4} \text{ or } \frac{3\pi}{4}\ ,\quad x = -\frac{\pi}{2} \text{ or } 0\ .$$

Next we let n be an even integer, say $n = 2m$, and write the equation in the form

$$\cos^{2m}x = 1 + \sin^{2m}x\ .$$

The left side never exceeds 1, while the right side exceeds 1 unless $\sin x = 0$; so the equation holds only if $x = 0$ or π.

Finally, let n be odd and greater than 1, say $n = 2m + 1$. Set $y = -x$. Then

$$\begin{aligned} 1 &= \cos^{2m+1}x - \sin^{2m+1}x = \cos^{2m+1}(-y) - \sin^{2m+1}(-y) \\ &= \cos^{2m+1}y + \sin^{2m+1}y \\ &\leqslant |\cos^{2m+1}y| + |\sin^{2m+1}y| \\ &= \cos^2 y|\cos^{2m-1}y| + \sin^2 y|\sin^{2m-1}y| \\ &\leqslant \cos^2 y + \sin^2 y \ = \ 1\ , \end{aligned}$$

Equality holds in the third line of this display if and only if $\cos y \geqslant 0$ and $\sin y \geqslant 0$, while equality holds in the last line if and only if $|\sin y| = 1$ or $|\cos y| = 1$. Therefore, either $\sin y = 1$ and $x = 2k\pi - \pi/2$, or $\cos y = 1$ and $x = 2k\pi$.

1961/4 First solution. Referring to Figure 11a and denoting the area of $\triangle XYZ$ by (XYZ), let

$$A = (PP_2P_3)\ , \qquad B = (PP_1P_3)\ , \qquad C = (PP_1P_2)\ .$$

Then $A + B + C = (P_1P_2P_3)$. Now triangles PP_2P_3 and $P_1P_2P_3$ have the same base P_2P_3, and their altitudes are in the ratio $PQ_1 : P_1Q_1$. Hence their areas are in this ratio, i.e. $PQ_1/P_1Q_1 = A/(A + B + C)$. Similarly

$$\frac{PQ_2}{P_2Q_2} = \frac{B}{(A+B+C)} \quad \text{and} \quad \frac{PQ_3}{P_3Q_3} = \frac{C}{(A+B+C)}\ .$$

Hence

$$\frac{PQ_1}{P_1Q_1} + \frac{PQ_2}{P_2Q_2} + \frac{PQ_3}{P_3Q_3} = 1\ ,$$

which implies that at least one of the ratios PQ_i/P_iQ_i is $\leqslant 1/3$, and one is $\geqslant 1/3$. This is equivalent to the proposed inequalities; for $PQ_i/P_iQ_i \leqslant 1/3$ if and only if

$$\frac{P_iQ_i}{PQ_i} = \frac{P_iP + PQ_i}{PQ_i} = \frac{P_iP}{PQ_i} + 1 \geqslant 3\ ,$$

and this holds if and only if $P_iP/PQ_i \geqslant 2$. Similarly, $PQ_i/P_iQ_i \geqslant 1/3$ if and only if $P_iP/PQ_i \leqslant 2$.

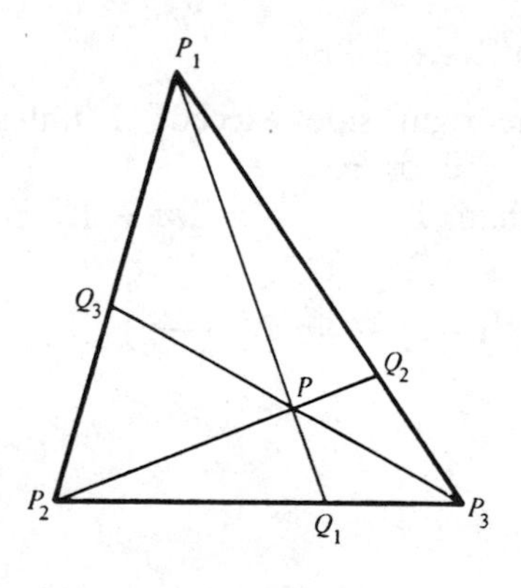

Figure 11a

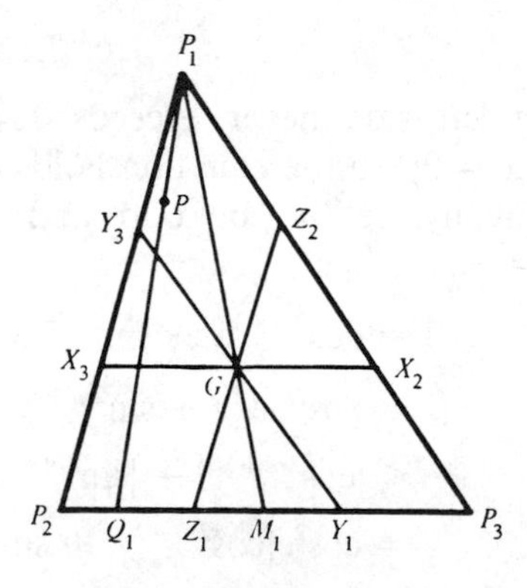

Figure 11b

Second solution. Through the centroid G of $\triangle P_1P_2P_3$, draw lines X_2X_3, Y_1Y_3, and Z_1Z_2 parallel to P_2P_3, P_1P_3 and P_1P_2 respectively (see Figure 11b). Since $P_1G/GM_1 = 2$, the inequality $P_1P/PQ_1 \leqslant 2$ holds if and only if P is in the triangle $P_1X_2X_3$. Similarly $P_2P/PQ_2 \leqslant 2$ if and only if P is in triangle $P_2Y_1Y_3$, and $P_3P/PQ_3 \leqslant 2$ if and only if P is in triangle $P_3Z_1Z_2$. Since the union of these triangles is all of $\triangle P_1P_2P_3$, at least one of the inequalities must hold. In the same way we see that one of the opposite inequalities must hold, since $\triangle P_1P_2P_3$ is the union of the trapezoids $P_2P_3X_2X_3$, $P_1P_3Y_1Y_3$, and $P_1P_2Z_1Z_2$.

1961/5. Our plan is to use side AC as chord of a circle whose major arc is the locus of points X such that $\angle AXC = \omega$; its minor arc is the locus of points Y with $\angle AYC = \pi - \omega$. Next we use the fact that a segment joining the midpoints of two sides of a triangle is half as long as the third side to locate the midpoint M of BC, $c/2$ units from the midpoint N of AC and on the minor arc of the circle. Finally we extend CM to locate B so that $CM = MB$.

Here are the details. Draw a segment AC of length b; denote its midpoint by N, and construct the perpendicular bisector of AC (see Figure 12a). With C as vertex and CN as one side, lay off $\angle NCX = (\pi - \omega)/2$, X being the point where the other side of that angle meets the perpendicular bisector of AC. Now construct the perpendicular bisector of CX, and let O be its intersection with NX. Draw a circle S_1 with center O and radius OC. With center N and radius $c/2$, draw a circle S_2. Suppose S_2 intersects S_1 in a point M. Then extend CM to B so that $MB = MC$. Triangle ABC is a solution of the problem.

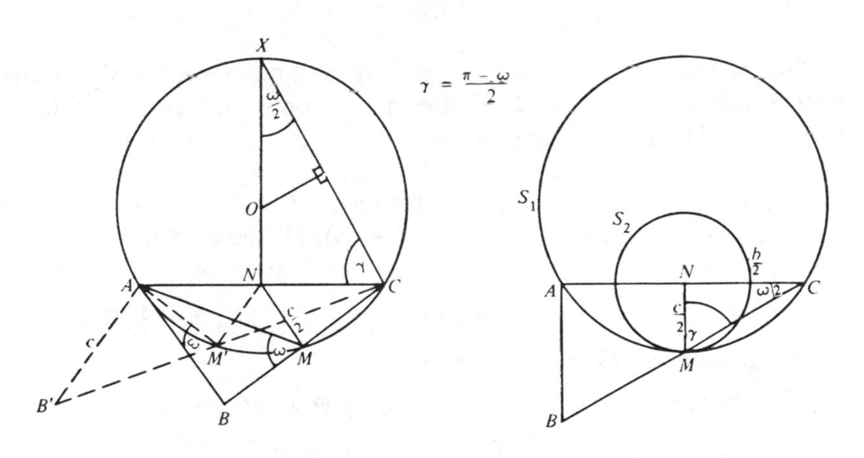

Figure 12a

Figure 12b

It remains to determine the conditions under which S_1 and S_2 actually intersect. If the radius $c/2$ is too small, S_2 lies entirely inside of S_1. As $c/2$ increases, there comes a moment when S_2 is tangent to S_1, as in Figure 12b. Then $\angle NMC = (\pi - \omega)/2$, so $\angle NCM = \omega/2$, and we have

$$\tan\frac{\omega}{2} = \frac{c/2}{b/2} = \frac{c}{b}, \quad \text{i.e.} \quad c = b\tan\frac{\omega}{2}.$$

In this case $\triangle ABC$ is a right triangle with legs b and c.

If $c > b\tan\frac{1}{2}\omega$, S_1 and S_2 intersect in two points M and M'. This gives two solutions, $\triangle ABD$ and $\triangle AB'C$, where $\angle BAC$ is acute and $\angle B'AC$ is obtuse (see Figure 12a).

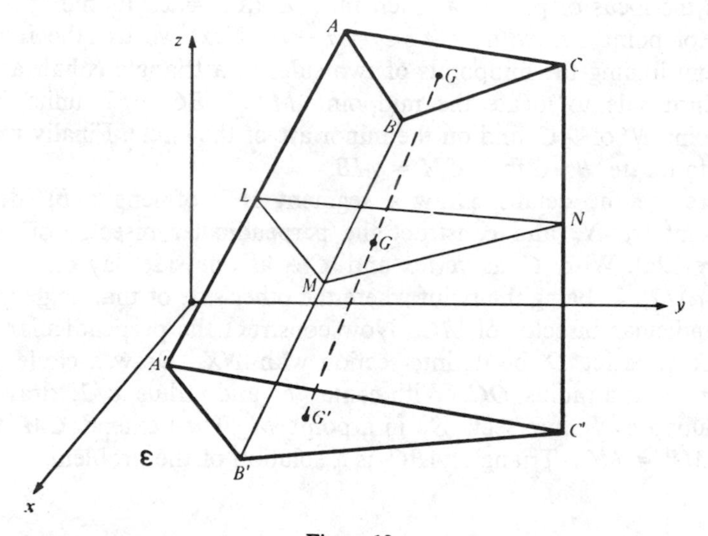

Figure 13

1961/6. Let ε be the plane $z = 0$ of a three-dimensional cartesian system, and let points A, B, C be in the upper half-space $z > 0$, see Figure 13. Define the sum of two points $X = (x_1, x_2, x_3)$ and $Y = (y_1, y_2, y_3)$ to be the point $X + Y = (x_1 + y_1, x_2 + y_2, x_3 + y_3)$. It is then easy to see that the midpoint of a segment XY is $(X + Y)/2$, and the centroid of a $\triangle XYZ$ is $(X + Y + Z)/3$; see also the solution of 1960/5, p. 32. The midpoints L, M, N of segments AA', BB', CC' are

$$L = \tfrac{1}{2}(A + A'), \qquad M = \tfrac{1}{2}(B + B'), \qquad N = \tfrac{1}{2}(C + C'),$$

hence the centroid G of $\triangle LMN$ is

$$\begin{aligned} G &= \tfrac{1}{3}(L + M + N) = \tfrac{1}{6}(A + A' + B + B' + C + C') \\ &= \tfrac{1}{2}\left[\tfrac{1}{3}(A + B + C) + \tfrac{1}{3}(A' + B' + C')\right] = \tfrac{1}{2}(\overline{G} + G'), \end{aligned}$$

where $\overline{G}$ is the centroid of the given triangle ABC, and G' is the centroid of the variable $\triangle A'B'C'$ in the plane $z = 0$. Hence the locus of G is the plane $z = \frac{1}{6}(a_3 + b_3 + c_3)$, parallel to $z = 0$ and half the distance from $\bar{G}$ to the plane $z = 0$. (Here a_3, b_3, c_3 are the z-coordinates of A, B, C.)

Fourth International Olympiad, 1962

1962/1. Suppose the desired number has $k + 1$ digits, and write it in the form $10N + 6$; then the transformed number is $6 \cdot 10^k + N$. The problem requires that

$$6 \cdot 10^k + N = 4(10N + 6) \ ,$$

which, when simplified, becomes

$$2 \cdot 10^k - 8 = 13N \ . \tag{1}$$

This equation tells us that the number on the left, which has digits $199 \cdots 2$ (where there are $k - 1$ 9's) is divisible by 13. We divide it by 13 to determine the quotient N:

```
      1 5 3 8 4
13 | 1 9 9 9 9 . . .2
     1 3
     ---
       6 9
       6 5
       ---
         4 9
         3 9
         ---
         1 0 9
         1 0 4
         -----
             5 2
             5 2
```

Thus the smallest possible value of N is 15384, and the desired number is $n = 153846$. If there were a smaller number with the required properties, we would have found a smaller exact quotient in the above division.

We could also have determined the number $k - 1$ of 9's in $13N = 199 \cdots 2$ as follows: From (1) we obtain

$$2 \cdot 10^k \equiv 8 (\text{mod } 13) \ ,$$

$$10^k \equiv 4 (\text{mod } 13)$$

$$10^{k+1} \equiv 40 \equiv 1 (\text{mod } 13) \ .\dagger$$

The task now is to find the smallest power $k + 1$ such that $10^{k+1} \equiv 1$ (mod 13). We find that $k + 1 = 6$, whence $k - 1 = 4$ and $13N = 199992$.

†See p. 195

The desired result can also be obtained by successive multiplication. Since the original number ends in 6, the new number ends in 4. Since the new number begins with 6 and ends in 4, the old number must have ended in 46. Now successive multiplications give in succession, the remaining digits of the original number. The process is as follows:

Original Number		New Number
. . . . 6 ×4	=	6 4
. . . . . . 4 6 ×4	=	6 . . . . 8 4
. . . . 8 4 6 ×4	=	6 . . . . . . 3 8 4
. . . 3 8 4 6 ×4	=	6 . . . . . 5 3 8 4
. 5 3 8 4 6 ×4	=	6 . . . . 1 5 3 8 4
1 5 3 8 4 6 ×4	=	6 1 5 3 8 4

1962/2. Set $f(x) = \sqrt{3-x} - \sqrt{x+1}$. In order for $f(x)$ to be real we must have $-1 \leqslant x \leqslant 3$. In this interval $f(x)$ decreases continuously from $f(-1) = 2$ to $f(3) = -2$. Hence there is a unique value $x = a$ such that $f(a) = \frac{1}{2}$, and the solution consists of all x such that $-1 \leqslant x < a$. We note that $f(1) = 0$, so $a < 1$. To find a we solve the equation $f(a) = \frac{1}{2}$:

$$\sqrt{3-a} - \sqrt{a+1} = \frac{1}{2} .$$

Squaring both sides, we find that

$$4 - 2\sqrt{(3-a)(a+1)} = 1/4 , \quad \text{and hence} \quad \sqrt{(3-a)(a+1)} = 15/8 .$$

Squaring again, we obtain the quadratic equation

$$a^2 - 2a + \frac{33}{64} = 0 .$$

The roots of this equation are $a = 1 \pm \sqrt{31}/8$. Since $a < 1$, we have $a = 1 - \sqrt{31}/8$, so the solution is

$$-1 \leqslant x < 1 - \sqrt{31}/8 .$$

(The root $1 + \sqrt{31}/8$ of the quadratic equation for a is an extraneous root introduced by the squaring operations; $f(1 + \sqrt{31}/8) = -1/2$.)

1962/3. Introduce a coordinate system, as shown in Figure 14. Let $0 \leqslant t \leqslant 1$ be the time interval during which the points X and Y traverse the first edges of their paths. Then the positions of X and Y at any time $t (0 \leqslant t \leqslant 4)$ are tabulated below. Here capital letters denote points and also vectors from the origin to these points; see the solution of 1960/5, p. 31 for definitions of our notation.

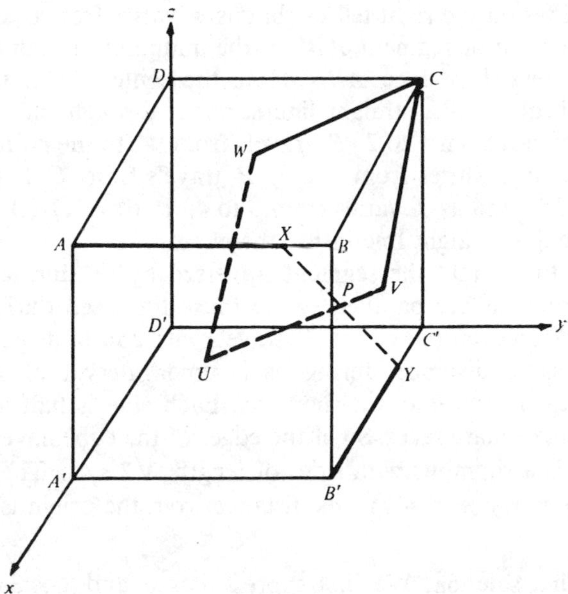

Figure 14

t	X	Y
$0 \leqslant t \leqslant 1$	$(1-t)A + tB$	$(1-t)B' + tC'$
$1 \leqslant t \leqslant 2$	$(2-t)B + (t-1)C$	$(2-t)C' + (t-1)C$
$2 \leqslant t \leqslant 3$	$(3-t)C + (t-2)D$	$(3-t)C + (t-2)B$
$3 \leqslant t \leqslant 4$	$(4-t)D + (t-3)A$	$(4-t)B + (t-3)B'$

The midpoint P of segment XY therefore occupies the following positions:

$$P = \frac{X+Y}{2} = \begin{cases} \frac{1}{2}\left[(1-t)(A+B') + t(B+C')\right] \\ \quad = (1-t)\dfrac{A+B'}{2} + t\dfrac{B+C'}{2}\,, & 0 \leqslant t \leqslant 1 \\ \frac{1}{2}\left[(2-t)(B+C') + (t-1)(2C)\right] \\ \quad = (2-t)\dfrac{B+C'}{2} + (t-1)C\,, & 1 \leqslant t \leqslant 2 \\ \frac{1}{2}\left[(3-t)2C + (t-2)(B+D)\right] \\ \quad = (3-t)C + (t-2)\dfrac{B+D}{2}\,, & 2 \leqslant t \leqslant 3 \\ \frac{1}{2}\left[(4-t)(B+D) + (t-3)(A+B')\right] \\ \quad = (4-t)\dfrac{B+D}{2} + (t-3)\dfrac{A+B'}{2}\,, & 3 \leqslant t \leqslant 4\,. \end{cases}$$

The expressions on the right tell us that as t varies from 0 to 1, P travels from the midpoint of segment AB' to the midpoint of segment BC'; i.e., from the center U of face $ABB'A'$ to the center V of face $BCC'B'$, and the path of P is a straight line segment through the interior of the cube. As t varies from 1 to 2, P travels from V to the corner C along a straight line. As t varies from 2 to 3, P travels from C to the center W of face $ABCD$, and as t varies from 3 to 4, P travels back to its starting point U along a straight line inside the cube.

It is easy to see that the segment traversed by P during the first and third time intervals are parallel, as are those traversed during the second and fourth. Moreover, the vector P travels with constant speed, and hence covers the same distance during each time interval of length 1. We conclude that P traces out a rhombus. Each side is half as long as the diagonal of the square faces. So, if the edges of the cube have length s, the locus of P is a rhombus with sides of length $\sqrt{2}\, s/2$. The rhombus lies in the plane $x + y + z = 2s$; its distance from the origin is $2s/\sqrt{3}$.

1962/4 First solution. We first express $\cos^2 x$ and $\cos^2 3x$ in terms of $\cos 2x$ and $\cos 6x$ by means of the half angle formula:

$$\cos^2 x = \tfrac{1}{2}(1 + \cos 2x), \quad \cos^2 3x = \tfrac{1}{2}(1 + \cos 6x) .$$

When these are substituted into the given equation, it becomes

$$\text{(1)} \qquad \cos 2x + 2\cos^2 2x + \cos 6x = 0 .$$

Now we recall a corollary of the addition formula for the cosine; $\cos(\alpha + \beta) + \cos(\alpha - \beta) = 2\cos\alpha\cos\beta$ and apply it with $\alpha = 4x$, $\beta = 2x$, obtaining

$$\cos 6x + \cos 2x = 2\cos 4x \cos 2x .$$

When this is substituted into (1), we get

$$\cos 2x[\cos 4x + \cos 2x] = 0 .$$

Another application of the above identity, with $\alpha = 3x$, $\beta = x$, yields

$$\cos 4x + \cos 2x = 2\cos 3x \cos x ,$$

so that (1) now becomes

$$\cos x \cos 2x \cos 3x = 0 .$$

This is satisfied when

$\cos x = 0$, i.e. $x = (2k+1)\dfrac{\pi}{2}$; or $\cos 2x = 0$, i.e. $x = (2k+1)\dfrac{\pi}{4}$;

or $\cos 3x = 0$, i.e. $x = (2k+1)\dfrac{\pi}{6}$; $\qquad k = 0, 1, 2 \ldots$.

Remark. The above solution may seem a bit tricky. One can also arrive at the solution in a more routine manner. Once the equation has been transformed into one involving $\cos 2x$, $\cos 4x$, $\cos 6x$, all we need to notice is that it can be written as a cubic in $\cos 2x$, which factors immediately so that its solutions can be read off.

Note. Another solution involves the use of complex numbers. We first give the necessary background material for readers who have not learned it so far.

We shall need the polar representation of complex numbers:

$$z = u + iv = r(\cos\theta + i\sin\theta)\ ,$$

where u and v are real, $i = \sqrt{-1}$, $r = \sqrt{u^2 + v^2}$, $\theta = \arctan(v/u)$. Our concern here is with complex numbers whose modulus r is 1, i.e. complex numbers of the form $z = \cos\theta + i\sin\theta$. Suppose that

$$z_1 = \cos\theta_1 + i\sin\theta_1 \qquad \text{and} \qquad z_2 = \cos\theta_2 + i\sin\theta_2$$

are two such numbers. Then

$$\text{(2)} \qquad \begin{aligned} z_1 z_2 &= (\cos\theta_1\cos\theta_2 - \sin\theta_1\sin\theta_2) + i(\sin\theta_1\cos\theta_2 + \sin\theta_2\cos\theta_1) \\ &= \cos(\theta_1 + \theta_2) + i\sin(\theta_1 + \theta_2) \end{aligned}$$

by the addition formulas for sine and cosine. More generally, if $z_1 = \cos\theta_1 + i\sin\theta_1, \cdots, z_n = \cos\theta_n + i\sin\theta_n$ are n complex numbers of modulus 1, then

$$\text{(2)}' \qquad z_1 z_2 \ldots z_n = \cos(\theta_1 + \theta_2 + \cdots + \theta_n) + i\sin(\theta_1 + \theta_2 + \cdots + \theta_n)\ .$$

This follows easily by mathematical induction on n.

If in (2′) we put $z_1 = z_2 = \cdots = z_n = z = \cos\theta + i\sin\theta$, we obtain de Moivre's theorem:

$$(\cos\theta + i\sin\theta)^n = \cos n\theta + i\sin n\theta\ .$$

Moreover, $(\cos\theta + i\sin\theta)(\cos\theta - i\sin\theta) = \cos^2\theta + \sin^2\theta = 1$, so that $\cos\theta - i\sin\theta$ is the reciprocal of $\cos\theta + i\sin\theta$. Thus if $z = \cos\theta + i\sin\theta$, then

$$z^n = \cos n\theta + i\sin n\theta \qquad \text{and} \qquad z^{-n} = \cos n\theta - i\sin n\theta\ .$$

Solving these two equations for $\cos n\theta$ and $i\sin n\theta$, we obtain

$$\text{(3)} \qquad \cos n\theta = \frac{z^n + z^{-n}}{2}\ , \qquad i\sin n\theta = \frac{z^n - z^{-n}}{2}\ .$$

Second solution. To conform to the notation just introduced, we replace the variable x in the problem by θ, writing the given equation as

$$\text{(1′)} \qquad \cos^2\theta + \cos^2 2\theta + \cos^2 3\theta = 1\ .$$

We make use of identity (3) above to write (1′) in the form

$$(z + z^{-1})^2 + (z^2 + z^{-2})^2 + (z^3 + z^{-3})^2 = 4\ ,$$

where $z = \cos\theta + i\sin\theta$ and $z^2 \neq 1$ since (1′) does not hold when θ is a multiple of π. After squaring the expressions on the left and obtaining

$$\text{(4)} \qquad z^2 + z^{-2} + z^4 + z^{-4} + z^6 + 2 + z^{-6} = 0\ ,$$

we rearrange the terms so that the left side is a geometric progression with common ratio $z^2 \neq 1$:

$$z^{-6} + z^{-4} + z^{-2} + 1 + z^2 + z^4 + z^6 = -1 .$$

We sum the terms on the left and get

$$(z^{-6} - z^8)/(1 - z^2) = -1 \qquad \text{or} \qquad z^7 - z^{-7} = -(z - z^{-1}) .$$

The last equation is equivalent, by (3), to $\sin 7\theta = -\sin\theta = \sin(-\theta)$; this is satisfied if

$$7\theta = \pi + \theta + 2k\pi , \qquad \theta = (1 + 2k)\pi/6 ,$$

and also if

$$7\theta = -\theta + 2k\pi , \qquad \theta = k\pi/4 , \quad k \not\equiv 0 \pmod 4) .$$

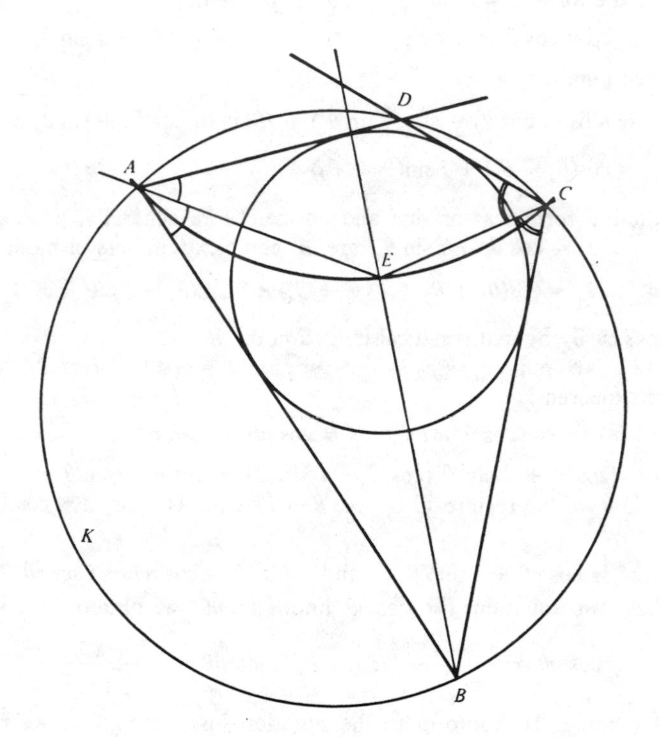

Figure 15

1962/5. We first locate the center E of the circle inscribed in the desired quadrilateral $ABCD$ by observing that E lies on the bisectors of angles A, B, and C, and that opposite angles (A and C) of a cyclic quadrilateral add up to 180°. We draw the bisector of $\angle B$; see Figure 15. We note that $\angle AEC$ is the sum of the exterior angles at E of triangles EAB and ECB. Since an exterior angle is the sum of the two opposite interior angles, we find that

$$\angle AEC = \angle EAB + \angle ECB + \angle ABC = \tfrac{1}{2}(\angle DAB + \angle DCB) + \angle ABC$$
$$= 90^\circ + \angle ABC\ .$$

The set of points P such that $\angle CPA = 90^\circ + \angle ABC$, and such that B is outside the angle is an arc of a circle, easy to construct (see e.g. 1961/5, p. 43). The intersection of this arc with the angle bisector of $\angle B$ is the point E.

We can now construct $A = 2\angle EAB$. One side of $\angle A$ is AB, the other intersects circle K in the desired point D.

Remark. We were not asked to prove that the required point D exists for any three given points A, B, C on K, yet we cannot consider our solution complete without looking into this question. It is, in fact, quite simple to prove that $ABCD$ has three concurrent angle bisectors, and that its opposite angles are supplementary (which guarantees the existence of the inscribed and circumscribed circles *provided* the point D given by our construction is such that $ABCD$ is a *simple* (i.e. non-self intersecting) quadrilateral.

Such questions of relative position are somewhat awkward to handle in non-analytic geometry. Only about a century ago was it fully realized that Euclid's proofs depend on tacit assumptions concerning which points lie between which others. To fill these gaps in Euclid's logic, additional axioms and arguments had to be introduced. This is not a matter of belaboring the obvious; there are "proofs" that all triangles are isosceles and other absurdities based on incorrect assumptions regarding the relative positions of certain points, and it is by no means obvious at first sight where the error lies.†

The following argument completes our solution of the problem. Let P be any point on arc AC of the circle ABC. The convex quadrilateral $ABCP$ has an inscribed circle if and only if

$$(1) \qquad AB + PC = AP + BC\ .$$

[This follows from the equality of tangent segments from a vertex to the inscribed circle.] Let P move from A to C along arc AC. When $P = A$, the left side of (1) is greater than the right, by the triangle inequality. When $P = C$, the triangle inequality tells us that the left side is smaller than the right. Since both sides vary continuously with P, there exists an intermediate position where equality holds. This position of P locates vertex D of our problem.

1962/6 First solution. In Figures 16, O is the circumcenter and I the incenter of isosceles $\triangle ABC$ with equal sides $AB = AC = s$.

There are three different cases, illustrated in Figures 16a, b, and c. In case (a), $A \leqslant 60^\circ$; in case (b), $60^\circ \leqslant A \leqslant 90^\circ$; and in case (c), $A \geqslant 90^\circ$. (In the borderline case $A = 60^\circ$, we have $O = I$; and when $A = 90^\circ$, O

† A good collection of such fallacies can be found in *Riddles in Mathematics* by E. P. Northrop, Pelican Books, 1960, Chapter 6.

is on BC.) We present the solution in the first case; it can be applied with minor modifications to the other two cases as well.

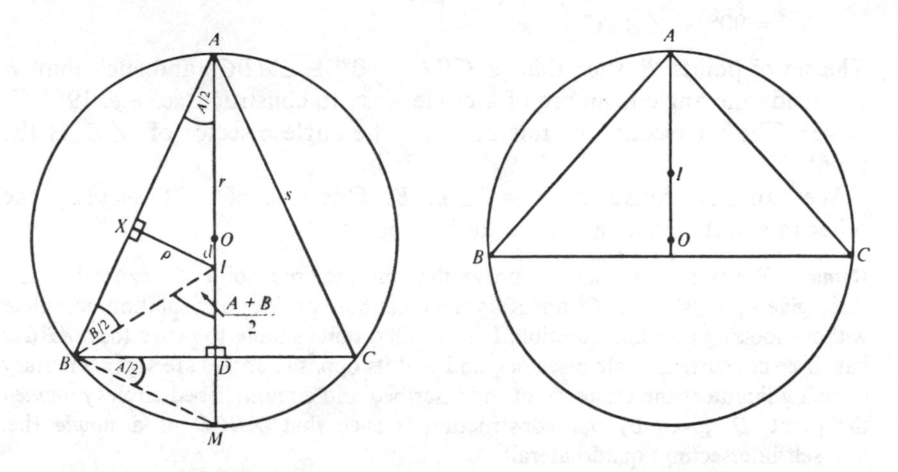

Figure 16a Figure 16b

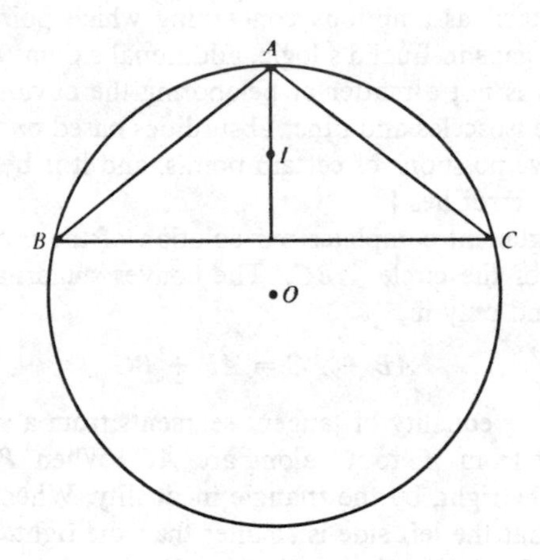

Figure 16c

In $\triangle ABI$, we have $\angle BAI = \frac{1}{2}A$ and $\angle ABI = \frac{1}{2}B$. Hence the exterior $\angle BIM = \frac{1}{2}(A + B)$. Also $\angle MBC = \frac{1}{2}A$ and $\angle IBC = \frac{1}{2}B$; so $\angle IBM = \frac{1}{2}(A + B)$. Thus $\triangle MBI$ is isosceles, with $BM = IM = r - d$.

Let X be the point where the incircle is tangent to side AB. The right triangles $\triangle AXI$ and $\triangle ABM$ are similar, so

$$\frac{IX}{BM} = \frac{AI}{AM}, \quad \text{or} \quad \frac{\rho}{r - d} = \frac{r + d}{2r}.$$

It follows that $r^2 - d^2 = 2r\rho$, so $d = \sqrt{r(r - 2\rho)}$.

Note. Euler proved that in *any* triangle, the distance d between the incenter and circumcenter satisfies

$$d^2 = r^2 - 2\rho r ,$$

where r and ρ are the radii of the circumcircle and incircle, respectively. For proofs, see e.g. *Geometry Revisted* by H. S. M. Coxeter and S. Greitzer, NML vol. 19 (1967), p. 29, *Geometric Inequalities* by N. D. Kazarinoff, NML vol. 4 (1961), p. 79, and *Hungarian Problem Book I*, NML vol. 11 in this series (1963), p. 51. A useful consequence of Euler's theorem is the inequality $r \geqslant 2\rho$, where equality holds if and only if the triangle is equilateral.

Since an isosceles triangle is determined by the size of the circumcircle and the incircle, it is obvious that every other quantity associated with the triangle, including d, can be expressed in terms of ρ and r. On the other hand there are many different scalene triangles with the same ρ and r; Euler's discovery was that the distance between the centers of the circles is the same for all of these. This leads to the remarkable fact that if two non-concentric circles are such that there is one triangle which is inscribed in one and circumscribed around the other, then there is a whole continuous family of noncongruent triangles with this property. Poncelet later generalized this: he showed that if two conics are such that there is an n-gon inscribed in one and circumscribed around the other, then there are infinitely many such n-gons.

Second solution (due to P. Herdeg). Let $h = AD$ be the altitude to BC. Then $h = r + d + \rho$. In $\triangle ADC$,

$$h = s \sin C . \tag{1}$$

In $\triangle ABM$,

$$\frac{AB}{AM} = \frac{s}{2r} = \sin \angle AMB = \sin C$$

(since both $\angle AMB$ and $\angle C$ are inscribed in the circular arc BCA). Hence

$$2r = \frac{s}{\sin C} . \tag{2}$$

In $\triangle BID$,

$$\rho = BD \tan \frac{B}{2} = BD \tan \frac{C}{2} .$$

Now $BD = DC = s \cos C$, so

$$\rho = s \cos C \tan \frac{C}{2} = s \cos C \frac{1 - \cos C}{\sin C} . \tag{3}$$

Solving the equation $h = r + d + \rho$ for d, and using (1), (2) and (3), we obtain

$$d = h - r - \rho = \frac{s}{2 \sin C} [2 \sin^2 C - 1 - 2 \cos C + 2 \cos^2 C]$$

$$= \frac{s}{2 \sin C} (1 - 2 \cos C) .$$

But

$$r^2 - 2r\rho = \frac{s^2}{4\sin^2 C} - \frac{s^2}{\sin C}\,\frac{\cos C(1-\cos C)}{\sin C}$$

$$= \frac{s^2}{4\sin^2 C}(1-2\cos C)^2 = d^2 .$$

It follows that $d = \sqrt{r^2 - 2r\rho}$.

1962/7. Solution (a) (due to G. Arenstorf). One of the five spheres touches all six edges internally; each of the other four touches the edges of one face internally, and the edges emanating from the opposite vertex externally.† Consider first the "inner" sphere Σ_0. Since the tangents to it from any vertex of the tetrahedron are equal, it follows (see Figure 17a) that

(1) $$SA + BC = SB + CA = SC + AB = x + y + z + w .$$

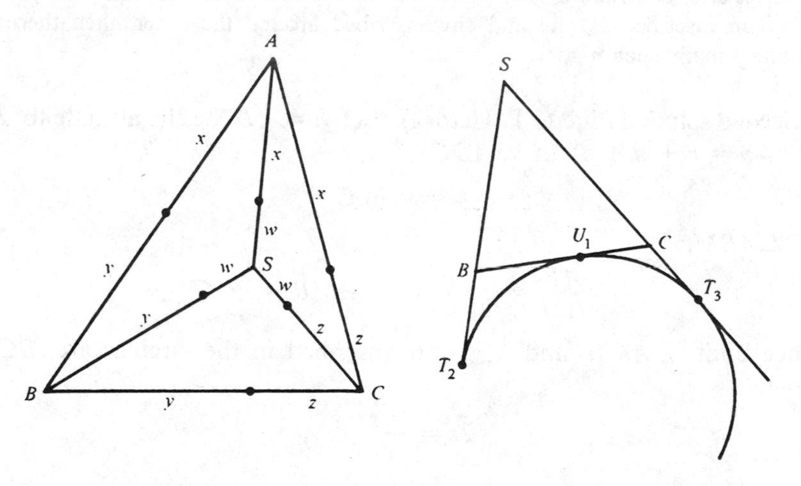

Figure 17a

Figure 17b

Next consider one of the "outer" spheres, say the one that touches BC, CA, AB internally and SA, SB, SC externally. Let T_1, T_2, T_3 be the points of tangency on SA, SB, SC extended, and let U_1, U_2, U_3 be the points of tangency on BC, CA, AB (see Figure 17b). Then

$$ST_1 = ST_2 = ST_3 .$$

Since

$$BU_1 = BT_2 \qquad \text{and} \qquad CU_1 = CT_3 ,$$

† Since this is not obvious, we supply a proof at the end of this problem.

the perimeter of $\triangle SBC$ is

$$SB + BU_1 + CU_1 + SC = ST_2 + ST_3 = 2ST_1 .$$

In the same way we find that all three of the triangles SBC, SCA, SAB have the same perimeter $2ST_1$. Since triangles SBC and SCA have the same perimeter and a common side SC, we have $SB + BC = SA + CA$. By (1) we also have $SB - BC = SA - CA$. Adding these relations, we obtain $SB = SA$. Similarly, $SC = SA$. By (1) it follows that $AB = BC = CA$.

So far we used Σ_0 and one other sphere to conclude that $\triangle ABC$ is equilateral and that the three other faces of $SABC$ are isosceles. Using any one of the remaining four spheres, we can infer the equality of all edges of the tetrahedron.

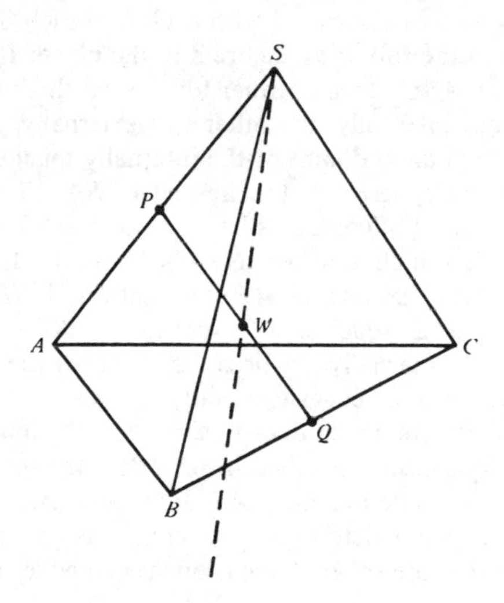

Figure 17c

(b) Conversely, suppose the tetrahedron $T = SABC$ is regular, and let W be its centroid, i.e. the center of mass of a system of four equal masses placed at the vertices. (By grouping these masses in pairs we see for example that W is the midpoint of PQ, where P and Q are the midpoints of SA and BC respectively; see Figure 17c.) Clearly W is invariant under any rotation which brings T into coincidence with itself. Since there exist such rotations carrying any edge of T onto any other edge, it follows that W is equidistant from the six edges. Hence there is a sphere with center W tangent to all these edges.

Now T is invariant under a 120° rotation about SW. It follows that any sphere Σ with center on SW extended which is tangent to SA is also tangent to SB and SC. Similarly, if Σ is tangent to AB it is also tangent to BC and CA. Thus we can construct an outer sphere by finding a point X on SW extended which is equidistant from SA and AB. The locus of points equidistant from SA and AB consists of two planes perpendicular to $\triangle SAB$ through the bisectors of $\angle SAB$. Of these planes, the one containing the internal bisector of $\angle SAB$ goes through W, while the other one intersects SW extended in the desired point X.

The construction of the other three outer spheres is completely analogous to this.

We now prove: If a sphere Σ touches all six edges of a tetrahedron, then one of the following alternatives holds:

(i) Σ touches all six edges internally

(ii) Σ touches the three edges of one face internally and the other three edges of the tetrahedron externally. Moreover, there is at most one sphere such that (i) holds and at most one sphere associated with each face such that (ii) holds.

Our proof is based on the following theorem in the plane: If a circle touches all three sides of a triangle ABC, then it either touches all three sides internally or it touches one of the sides internally, the other two externally, and each externally touched side goes through an endpoint of the internally touched side.

Now suppose that the sphere Σ touches edge SA of tetrahedron $SABC$ externally, and that the point of contact is beyond point A. Then Σ intersects the plane SAB in a circle which touches SA beyond A. Hence by the above theorem applied to $\triangle SAB, \Sigma$ touches AB internally and SB externally beyond B. Similarly we find that Σ touches AC internally, SC externally beyond C, and that it touches BC internally. In brief, Σ touches the three edges of face ABC internally, the other three edges externally.

There is at most one sphere which does this. Indeed, the intersection of Σ with plane ABC is the unique circle inscribed in $\triangle ABC$; the intersection of Σ with plane SAB is the unique circle touching AB internally, SA and SB externally. Since four non-coplanar points determine a sphere, Σ is uniquely determined. The same argument shows that the sphere which touches all edges internally, if there is one, is unique.

Thus if there are five spheres touching all edges, then one of them, say Σ_0, touches all edges internally while the other four each touch one triangle internally, the other three edges externally.

To show that all edges of $SABC$ are equal, we actually need only the existence of Σ_0 and two of the remaining spheres; see the last paragraph in the solution of part (a) of this problem.

Fifth International Olympiad, 1963

1963/1. If $p < 0$, we have

$$\sqrt{x^2 - p} + 2\sqrt{x^2 - 1} \geqslant \sqrt{x^2 - p} > x\ .$$

So in order for the equation to have a solution, we must have $p \geqslant 0$. Now write it in the form

$$2\sqrt{x^2-1} = x - \sqrt{x^2-p}$$

and square, obtaining

$$2x^2 + p - 4 = -2x\sqrt{x^2-p} \ .$$

Squaring again, and solving for x^2, we get

$$x^2 = \frac{(p-4)^2}{8(2-p)} \ .$$

Hence in order for a solution to exist, we must have $0 \leqslant p < 2$, and then the only possible solution is $x = (4-p)/\sqrt{8(2-p)}$. We now substitute this value of x into the original equation. After multiplying all terms by $\sqrt{8(2-p)}$, we obtain

$$|3p-4| + 2p = 4 - p \ ; \qquad \text{i.e.} \qquad |3p-4| = -(3p-4) \ .$$

Clearly this holds if and only if $3p - 4 \leqslant 0$, i.e. $p \leqslant 4/3$. Therefore the equation has solutions only when $0 \leqslant p \leqslant 4/3$, and then $x = (4-p)/\sqrt{8(2-p)}$.

1963/2. The locus depends on the position of point A relative to segment BC.

(a) If A belongs to the segment BC, then all points inside or on the sphere with diameter AB and all points inside or on the sphere with diameter AC belong to the locus; see Figure 18a. If A coincides with either endpoint of the segment, then one of the spheres shrinks to a point.

(b) If A lies on the extension of segment BC, say with B between A and C as in Figure 18b, then the locus consists of all points inside or on

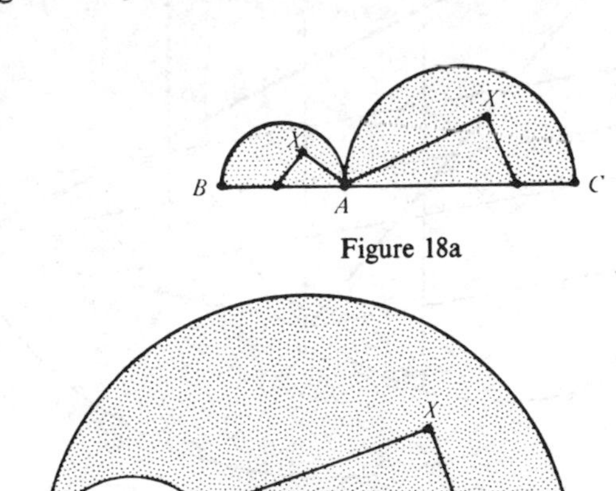

Figure 18a

Figure 18b

the sphere with diameter AC except those inside the sphere with diameter AB. For if Z is a point in this inner sphere, the plane through Z perpendicular to AZ cuts the line AC between A and B, and so does not intersect segment BC. Any point X in the shaded region clearly belongs to the locus.

(c) Suppose A does not lie on the line BC; see Figure 18c. Again we examine the spheres S_1 and S_2 with diameters AB and AC, respectively. Any point Y on the surface of S_1 (or S_2) is in the locus because $\angle AYB$ (resp. $\angle AYC$) is a right angle. Any point X in the interior of either S_1 or S_2 but not in the interior of both, is in the locus, because AX extended meets S_1 (resp. S_2) in a point Y_1 (resp. Y_2) such that $\angle AY_1B$ (resp. $\angle AY_2C$) is a right angle, and the plane through X perpendicular to AX intersects the segment BC. In contrast, any point Z in the region interior to both spheres does *not* belong to the locus. For AZ extended meets S_1 (resp. S_2) in a point W_1 (resp. W_2) such that $\angle AW_1B$ (resp. $\angle AW_2C$) is $90°$, but the plane through Z perpendicular to AZ does *not* intersect the segment BC (it intersects an extension of the segment BC instead).

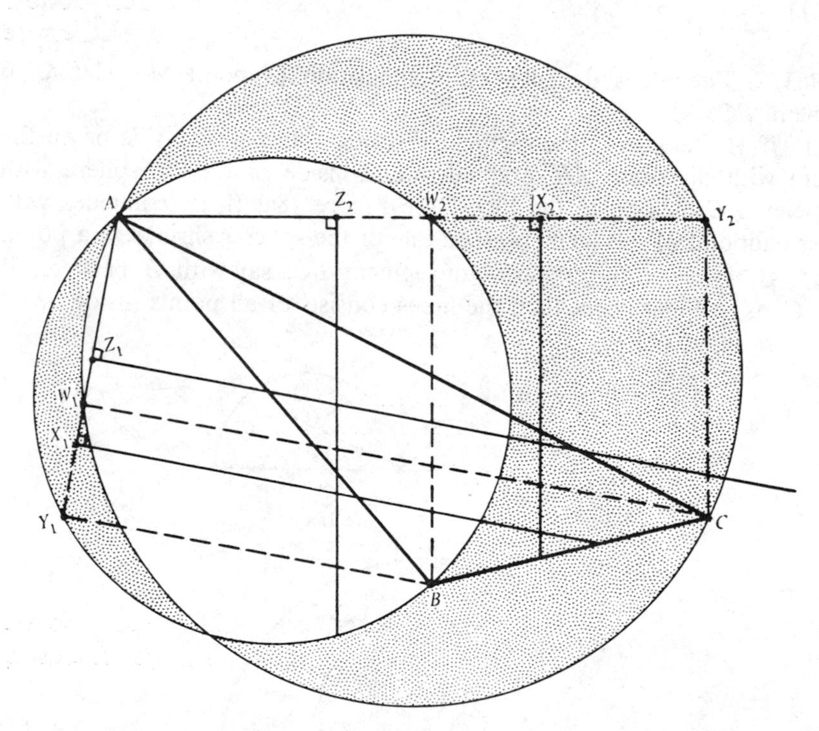

Figure 18c

Summarizing, we can describe the required locus as the union $\overline{K}_1 \cup \overline{K}_2$ minus the intersection $K_1 \cap K_2$, where K_1, K_2 denote the interior points of the spheres S_1, S_2, and $\bar{K}$ denotes the point set K together with its boundary S. [Strictly speaking, if we do not admit right angles AXQ where one of the sides, say XQ, has zero length, then we should exclude the segment BC from the loci shown in Figures 18a, b, c.]

1963/3 First solution. We will actually prove the conclusion assuming that only $n-1$ sides of the n-gon satisfy the given relation, i.e.

$$a_1 \geqslant a_2 \geqslant \cdots \geqslant a_{n-1} .$$

Let $P_1, P_2, \ldots, P_n$ be the vertices of the given polygon Π, where P_iP_{i+1} has length a_i. Suppose without loss of generality that P_nP_1 is horizontal. For each i ($i = 1, 2, \ldots, n-1$), let Π_i be the regular n-gon with side P_iP_{i+1} whose bottom edge is horizontal (see Figure 19, where $n = 6$). Since $a_i \geqslant a_{i+1}$, each polygon Π_i contains the next polygon Π_{i+1}. Whenever a strict inequality $a_i > a_{i+1}$ occurs, the bottom edge of Π_{i+1} is higher than the bottom edge of Π_i. So if *any* of the inequalities $a_1 > a_2$, $a_2 > a_3, \ldots, a_{n-2} > a_{n-1}$ holds, the bottom edge of Π_{n-1} is higher than that of Π_1. This is a contradiction, since P_n is on both these bottom edges. Hence $a_1 = a_2 = \cdots = a_{n-1}$. Clearly also $a_n = a_1$, for otherwise Π would not close at P_1.

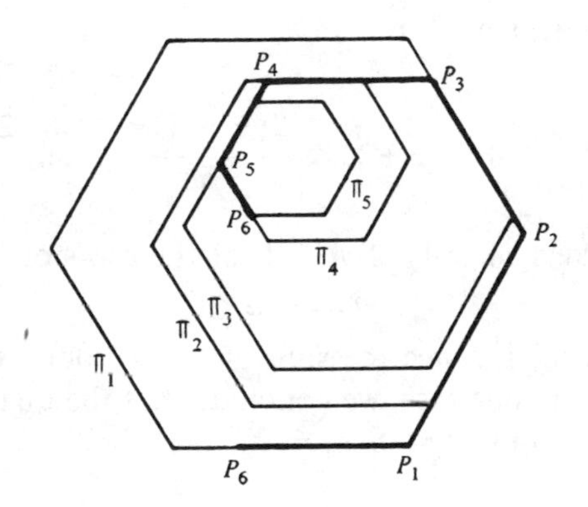

Figure 19

Second solution. Since the interior angles of the n-gon are equal, so are the exterior angles, each having measure $2\pi/n$. Since the n-gon is closed, we see that the vector sum

$$(1)\quad a_1(\cos 0 + i \sin 0) + a_2\left(\cos \frac{2\pi}{n} + i \sin \frac{2\pi}{n}\right) + \cdots$$
$$+ a_n\left(\cos \frac{2(n-1)\pi}{n} + i \sin \frac{2(n-1)\pi}{n}\right) = 0 .$$

Each, the real part and the imaginary part of (1), is 0; using the imaginary part and the fact that $\sin 0 = 0$, we get

$$(2) \qquad a_2 \sin\frac{2\pi}{n} + a_3 \sin\frac{4\pi}{n} + \ldots + a_n \sin\frac{2(n-1)\pi}{n} = 0 .$$

This equation has $n-1$ terms, and if we compare the $k-$ th and the $(n-k)-$ th we find that they contain the factors

$$\sin\frac{2k\pi}{n} \quad \text{and} \quad \sin\frac{2(n-k)\pi}{n} = \sin\left(2\pi - \frac{2k\pi}{n}\right) = -\sin\frac{2k\pi}{n} ,$$

respectively. We combine these pairs of terms and write (2) in the form

$$(3) \qquad (a_2 - a_n)\sin\frac{2\pi}{n} + (a_3 - a_{n-1})\sin\frac{4\pi}{n} + \ldots = 0 .$$

If n is odd, all terms of (2) are paired; if n is even, there is a middle term left over, namely $a_{(n+2)/2} \sin\pi = 0$. In any case, all terms of (3) are non-negative because all angles are between 0 and π, and all coefficients $a_j - a_m$ have $j < m$. It follows that each term of (3) is 0. This implies in particular that $a_2 = a_n$ which, together with the given inequalities, shows that $a_2 = a_3 = \ldots = a_n$. Thus (1) becomes

$$a_1 + a_2\left[\left(\cos\frac{2\pi}{n} + i\sin\frac{2\pi}{n}\right) + \ldots + \left(\cos\frac{2(n-1)\pi}{n} + i\sin\frac{2(n-1)\pi}{n}\right)\right] = 0 .$$

The sum in brackets consists of $n-1$ of the roots of

$$x^n - 1 = 0 ,$$

the n-th root being 1. Since the sum of all n roots is the coefficient of x^{n-1}, which is 0 in our case, we conclude that the quantity in brackets is -1, so that $a_1 + a_2(-1) = 0$, $a_1 = a_2$.

1963/4 First solution. An obvious solution is the so-called "trivial solution" $x_1 = x_2 = x_3 = x_4 = x_5 = 0$.

Adding the five given equations leads to

$$(x_1 + x_2 + x_3 + x_4 + x_5)(y - 2) = 0 ,$$

which implies that either $x_1 + x_2 + x_3 + x_4 + x_5 = 0$ or $y = 2$. If $y = 2$, (2) and (3) yield $x_1 + x_4 = x_2 + x_3$; and by (5), $x_1 + x_4 = 2x_5$. So

$$x_1 + x_2 + x_3 + x_4 + x_5 = 5x_5 .$$

Now the fact that the cyclic permutation $x_1 \to x_2 \to x_3 \to x_4 \to x_5 \to x_1$ does not alter the given system allows us to conclude that also

$$x_1 + x_2 + x_3 + x_4 + x_5 = 5x_i \qquad \text{for any } i \ .$$

It follows that all the x_i are equal. Conversely it is trivial to check that when $y = 2$, a solution is obtained by setting all the x_i equal to any fixed number.

To handle the case where $y \neq 2$, we make use of a systematic elimination process. Let us eliminate x_5 from the given equations, x_4 from the remaining four and x_3 from the remaining three; we find

$$(6) \qquad (y^2 + y - 1)(x_2 - x_1) = 0 \ ,$$

and

$$(7) \qquad (y^2 + y - 1)[x_2 - (y - 1)x_1] = 0 \ .$$

If $y^2 + y - 1 \neq 0$, then $x_1 = x_2$ by (6), and hence $(y - 2)x_1 = 0$ by (7). Since we are assuming $y \neq 2$, this implies that $x_1 = x_2 = 0$. The original equations then yield $x_3 = x_4 = x_5 = 0$, so we have the trivial solution mentioned earlier.

If $y^2 + y - 1 = 0$, (6) and (7) are satisfied for all values of x_1 and x_2. So x_1 and x_2 may be given arbitrary values; but by (2) $x_3 = yx_2 - x_1$, by (3) $x_4 = yx_3 - x_2 = (y^2 - 1)x_2 - yx_1$, and by (1) $x_5 = yx_1 - x_2$, where y is either root of the equation $y^2 + y - 1 = 0$.

This problem furnishes an excellent opportunity to illustrate some important and useful ideas of linear algebra. Those readers who have studied some vectors and matrices should have no difficulty following the solution given below.

Second solution. We can write all five given equations in the form

$$(8) \qquad x_{n+1} = -x_{n-1} + yx_n \ ,$$

provided we define $x_{k+5} = x_k$, $k = 0, 1, 2, \ldots$. Equation (8) is called a second order linear recursion formula.† To solve it, we write it in the language of 2×2 matrices. Define the matrix T and the vector v_n by

$$(9) \qquad T = \begin{pmatrix} 0 & 1 \\ -1 & y \end{pmatrix}, \qquad v_n = \begin{pmatrix} x_n \\ x_{n+1} \end{pmatrix}, \qquad n = 1, 2, \ldots \ .$$

In particular,

$$v_5 = \begin{pmatrix} x_5 \\ x_6 \end{pmatrix} = \begin{pmatrix} x_0 \\ x_1 \end{pmatrix} = v_0 \ .$$

Now (8) can be written in the form

$$v_{n+1} = Tv_n \ , \qquad n = 0, 1, 2, 3, 4 \ .$$

†The reader may have met similar recurrences in connection with the Fibonacci sequence.

It follows that

$$v_0 = v_5 = Tv_4 = T(Tv_3) = T^2v_3 = T^3v_2 = T^4v_1 = T^5v_0 \ . \tag{10}$$

We recall that a vector v is called an *eigenvector* of a matrix M if Mv is just a multiple of v: $Mv = \lambda v$, where the number λ is called an *eigenvalue* of M. So (10) says that v_0 is an eigenvector and 1 and eigenvalue of the matrix T^5.

There is an important relation between the eigenvectors and eigenvalues of a matrix and its powers: If v_0 is an eigenvector of T with eigenvalue t, then v_0 is also an eigenvector of T^k with eigenvalue t^k; for $Tv_0 = tv_0$ implies $T^kv_0 = T^{k-1}(Tv_0) = tT^{k-1}v_0 = \cdots = t^kv_0$. We shall use a partial converse of this without proof; that all eigenvalues of T^k are of the form t^k, where t is an eigenvalue of T. So we conclude that 1 is an eigenvalue of T^5 if and only if T has an eigenvalue t whose fifth power is 1:

$$t^5 = 1 \ . \tag{11}$$

The eigenvalues t of T satisfy $Tv = tv$, or $(T - tI)v = 0$ where v is the corresponding eigenvector, $I = \binom{10}{01}$ is the identity matrix, and the matrix $T - tI$ can map v into the zero vector only if its determinant vanishes. This leads to the so-called *characteristic equation*† for the eigenvalues,

$$\det(T - tI) = \begin{vmatrix} -t & 1 \\ -1 & y - t \end{vmatrix} = t^2 - yt + 1 = 0 \ . \tag{12}$$

This equation is satisfied for

$$y = t + \frac{1}{t} \ , \tag{13}$$

where t satisfies (11). The solutions of (11) are

$$t = \cos\frac{2\pi m}{5} + i\sin\frac{2\pi m}{5} \ , \qquad m = 0, 1, 2, 3, 4 \ .$$

For $m = 0, t = 1$; so $y = 2$, and all the x_i are equal (see first solution).

For $m \neq 0$, we can either substitute t into (13) or proceed as follows. Since (11) is equivalent to

$$t^5 - 1 = (t - 1)(t^4 + t^3 + t^2 + t + 1) = 0 \ ,$$

we have $t^4 + t^3 + t^2 + t + 1 = 0$. Using (13) we compute

$$y^2 + y = \left(t^2 + 2 + \frac{1}{t^2}\right) + \left(t + \frac{1}{t}\right)$$

and by virtue of (11), we can write it in the form

$$y^2 + y = t^2 + 2 + t^3 + t + t^4 = t^4 + t^3 + t^2 + t + 1 + 1 = 1 \ .$$

This leads to the quadratic equation

$$y^2 + y - 1 = 0$$

for y, which we met in the first solution, and to the corresponding solution sets x_1, x_2, x_3, x_4, x_5 described there.

†Note that any linear k-th order recursion formula $u_{j+1} = \sum_{i=j-k+1}^{j} c_i u_i$ can be treated in an analogous manner with a $k \times k$ matrix; the eigenvalues of the matrix lead to solutions of the recursion relation.

1963/5 First solution. The roots of the polynomial $f(x) = x^7 - 1$ are the seven complex numbers

$$\cos\frac{2\pi k}{7} + i\sin\frac{2\pi k}{7} \qquad (k = 0, 1, 2, \ldots, 6)\ .\dagger$$

Their sum is zero since the coefficient of x^6 in $f(x)$ is zero. Therefore the sum of their real parts is zero:

$$1 + \cos\frac{2\pi}{7} + \cos\frac{4\pi}{7} + \cos\frac{6\pi}{7} + \cos\frac{8\pi}{7} + \cos\frac{10\pi}{7} + \cos\frac{12\pi}{7} = 0\ .$$

Since

$$(1) \qquad \cos\theta = \cos(2\pi - \theta) = -\cos(\pi - \theta) = -\cos(\theta - \pi)\ ,$$

this equation may be rewritten as

$$1 + \cos\frac{2\pi}{7} - \cos\frac{3\pi}{7} - \cos\frac{\pi}{7} - \cos\frac{\pi}{7} - \cos\frac{3\pi}{7} + \cos\frac{2\pi}{7} = 0\ .$$

Hence $2(\cos\frac{\pi}{7} - \cos\frac{2\pi}{7} + \cos\frac{3\pi}{7}) = 1$, so

$$\cos\frac{\pi}{7} - \cos\frac{2\pi}{7} + \cos\frac{3\pi}{7} = \frac{1}{2}\ .$$

Second solution. We will prove the desired equality in the equivalent form

$$(2) \qquad \cos\frac{\pi}{7} + \cos\frac{5\pi}{7} + \cos\frac{3\pi}{7} = \frac{1}{2}$$

using the identity

$$(3) \qquad 2\cos x\sin y = \sin(x + y) - \sin(x - y)\ .$$

Applying (3) with $(x, y) = (\pi/7, \pi/7)$, $(3\pi/7, \pi/7)$, and $(5\pi/7, \pi/7)$, we obtain

$$2\cos\frac{\pi}{7}\sin\frac{\pi}{7} = \sin\frac{2\pi}{7} - \sin 0$$

$$(4) \qquad 2\cos\frac{3\pi}{7}\sin\frac{\pi}{7} = \sin\frac{4\pi}{7} - \sin\frac{2\pi}{7}$$

$$2\cos\frac{5\pi}{7}\sin\frac{\pi}{7} = \sin\frac{6\pi}{7} - \sin\frac{4\pi}{7} = \sin\frac{\pi}{7} - \sin\frac{4\pi}{7}\ .$$

Adding these three equations, we get

$$2\sin\frac{\pi}{7}\left(\cos\frac{\pi}{7} + \cos\frac{3\pi}{7} + \cos\frac{5\pi}{7}\right) = \sin\frac{\pi}{7}\ ,$$

and dividing by $2\sin(\pi/7)$ yields equations (2).

†This is a consequence of de Moivre's theorem, see the note following the first solution of 1962/4.

The same result could have been attained by means of variants of identity (3) leading to the common factor $\cos(\pi/7)$ on the left of the analogue of (4).

Note. Interchanging the second and third terms in (2) leads to a sum of cosines whose arguments form an arithmetic progression. This sum can be regarded as the real part of a geometric progression (see the note and second solution of 1962/4), and summing this progression leads to yet another solution of 1963/5.

1963/6. In the order DAECB of the second prediction, there are four consecutive pairs (DA), (AE), (EC) and (CB). Disjointness reduces the successful predictions to (i) (DA), (EC) B; (ii) (DA) (CB) E; (iii) (AE) (CB) D. Each of these triples of symbols is capable of six arrangements. Writing out all eighteen, we easily find that:

in (i), only (DA)B(EC) has exactly two contestants in the predicted order;

in (ii), only E(DA)(CB) and (DA)(CB)E have exactly two contestants in the predicted order;

in (iii), only (AE)D(CB) has exactly two contestants in the predicted order.

Now by what we know about the first prediction, we may eliminate DABEC, DACBE, and AEDCB, leaving EDACB as the actual order.

Sixth International Olympiad, 1964

1964/1. (a) We look for powers of 2 congruent to 1 modulo 7, and find that $2^1 \equiv 2 \pmod 7$, $2^2 \equiv 4 \pmod 7$, while $2^3 \equiv 1 \pmod 7$.† It follows that for every natural number k, $2^{3k} = (2^3)^k \equiv 1^k \equiv 1 \pmod 7$. Hence, every number of the form $2^{3k} - 1$ is divisible by 7. When n is not a multiple of three, it is of the form $3k + 1$ or $3k + 2$. Since $2^{3k} \equiv 1 \pmod 7$, we have

$$(1) \quad 2^{3k+1} = 2 \cdot 2^{3k} \equiv 2 \pmod 7, \qquad 2^{3k+2} = 4 \cdot 2^{3k} \equiv 4 \pmod 7$$

from which it follows that multiples of 3 are the only exponents n such that $2^n - 1$ is divisible by 7.

(b) From relations (1) it follows that

$$2^{3k+1} + 1 \equiv 3 \pmod 7, \qquad 2^{3k+2} + 1 \equiv 5 \pmod 7;$$

moreover $2^{3k} + 1 \equiv 2 \pmod 7$. Thus $2^n + 1$ leaves a remainder of 2, 3 or 5 when divided by 7, and hence is not divisible by 7.

1964/2 First solution (due to G. Arenstorf and A. Zisook). Let

$$x = b + c - a, \qquad y = c + a - b, \qquad z = a + b - c;$$

†See p. 197

note that x, y, z are positive, since a, b, c are sides of a triangle. Then

$$\frac{z+y}{2}=c\ , \qquad \frac{y+z}{2}=a\ , \qquad \frac{z+x}{2}=b\ .$$

According to the arithmetic mean-geometric mean inequality,†

$$\frac{x+y}{2}\geqslant\sqrt{xy}\ , \qquad \frac{y+z}{2}\geqslant\sqrt{yz}\ , \qquad \frac{z+x}{2}\geqslant\sqrt{zx}\ .$$

Therefore

$$\frac{x+y}{2}\cdot\frac{y+z}{2}\cdot\frac{z+x}{2}\geqslant xyz\ ,$$

which, on substitution, yields

$$abc\geqslant(b+c-a)(c+a-b)(a+b-c)\ .$$

When the factors on the right are multiplied out, the terms may be arranged so that this inequality takes the form

$$a^2(b+c-a)+b^2(c+a-b)+c^2(a+b-c)-2abc\leqslant abc\ ,$$

from which we arrive at the desired inequality.

Second solution (due to D. Barton). Label the sides so that $0\leqslant a\leqslant b\leqslant c$. Then

$$c-a\geqslant b-a\geqslant 0\ ,$$

and

$$c(c-b)(c-a)\geqslant b(c-b)(b-a)\geqslant 0$$

or

$$c(c-b)(c-a)-b(c-b)(b-a)\geqslant 0\ .$$

We add the non-negative term $a(a-b)(a-c)$ to the left side, obtaining

$$a(a-b)(a-c)+c(c-b)(c-a)-b(c-b)(b-a)\geqslant 0\ ,$$

or, equivalently,

$$a^3+b^3+c^3-a^2b-a^2c-c^2b-c^2a-b^2c-b^2a+3abc\geqslant 0\ .$$

This may be written in the form

$$a^2(a-b-c)+b^2(b-c-a)+c^2(c-b-a)+3abc\geqslant 0\ .$$

Multiplying by -1 and transposing the last term to the right, we obtain

$$a^2(b+c-a)+b^2(c+a-b)+c^2(a+b-c)\leqslant 3abc\ .$$

Equality holds if and only if $a=b=c$.

Third solution (due to C. Hornig). The left side of the inequality to be proved may be written in the form

$$a(b^2+c^2-a^2)+b(c^2+a^2-b^2)+c(a^2+b^2-c^2)\ .$$

†See p. 196

By the law of cosines, this expression is equal to $a(2bc \cos A) + b(2ac \cos B) + c(2ab \cos C) = 2abc(\cos A + \cos B + \cos C)$, where $A + B + C = 180°$. Now if we keep C fixed, the expression $\cos A + \cos B + \cos C$ is greatest when $A = B$; for

$$\cos A + \cos B = \cos \frac{A+B}{2} \cos \frac{A-B}{2} ,$$

and since $A + B$ is constant, this product is greatest when $\cos \frac{1}{2}(A - B) = 1$; i.e. for $A = B$. Since A, B, C enter the problem symmetrically, $\cos A + \cos B + \cos C$ has its maximum value when $A = B = C = 60°$. But then

$$\cos A + \cos B + \cos C = 3/2 ,$$

so

$$a^2(b + c - a) + b^2(c + a - b) + c^2(a + b - c) \leqslant 2abc\left(\tfrac{3}{2}\right) = 3abc .$$

1964/3. Consider one of the cut-off triangles, say $\triangle APQ$, where PQ is parallel to BC; see Figure 20. Since $\triangle APQ \sim \triangle ABC$, the inradius r_a of $\triangle APQ$ has the same ratio to the inradius r of $\triangle ABC$ as any segment in $\triangle APQ$ has to the corresponding segment of $\triangle ABC$. In particular, the altitude from A of $\triangle APQ$, which differs from the corresponding altitude h of $\triangle ABC$ by the diameter $2r$ of its inscribed circle, is in that ratio to h. Thus

$$\frac{h-2r}{h} = \frac{r_a}{r} = 1 - \frac{2r}{h} .$$

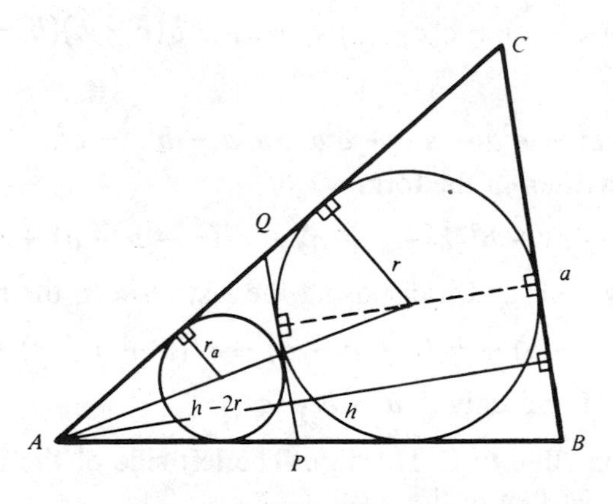

Figure 20

Let (ABC) denote the area of triangle ABC. Since $(ABC) = ah/2$, we have $h = 2(ABC)/a$, and hence

$$\frac{r_a}{r} = 1 - \frac{ra}{(ABC)} .$$

In the same way, we find that the radii r_b, r_c of the incircles of the other cut-off triangles satisfy

$$\frac{r_b}{r} = 1 - \frac{rb}{(ABC)} \quad \text{and} \quad \frac{r_c}{r} = 1 - \frac{rc}{(ABC)} .$$

Squaring each equality and adding, we find

$$(1) \qquad \frac{r_a^2 + r_b^2 + r_c^2}{r^2} = 3 - \frac{2r(a+b+c)}{(ABC)} + \frac{r^2(a^2+b^2+c^2)}{(ABC)^2}$$

Now

$$(2) \qquad (ABC) = \frac{1}{2}ar + \frac{1}{2}br + \frac{1}{2}cr = r\frac{a+b+c}{2} = rs ,$$

where $s = \dfrac{a+b+c}{2}$.

Hence the second term on the right side of (1) is equal to 4, and we can write (1) in the form

$$\frac{r_a^2 + r_b^2 + r_c^2}{r^2} = \frac{r^2(a^2+b^2+c^2)}{(ABC)^2} - 1 ,$$

or

$$(3) \qquad \frac{r_a^2 + r_b^2 + r_c^2 + r^2}{r^2} = \frac{r^2(a^2+b^2+c^2)}{(ABC)^2} .$$

In (3) we substitute $r = (ABC)/s$ from (2) and use Heron's formula to write $(ABC)^2 = r^2s^2 = s(s-a)(s-b)(s-c)$. This yields

$$r_a^2 + r_b^2 + r_c^2 + r^2 = (a^2+b^2+c^2)(s-a)(s-b)(s-c)/s^3 ;$$

the desired area is π times this quantity.

1964/4. Select any person, say A. He corresponds with 16 other people. Since there are only three topics, he must write to at least six of them on the same topic. This is an example of the "pigeonhole principle."†

Assume for definiteness that A writes to at least 6 of the other 16 people on topic I. If any one of these six writes to another of them on topic I, then there are 3 writers corresponding on topic I. So assume that these

†See p. 201

six write each other only on topics II and III. If B is one of the six, then by the pigeonhole principle, he must write to at least 3 of the other 5 on one of the two topics, say II.

Now there are two possibilities for these last three people. If one writes to another on topic II, then we have found three people corresponding on topic II. If, on the other hand, none of the three writes to another on topic II, then all three must write to each other on topic III. This proves the assertion.

Remark. A beautiful theorem of Ramsey says the following. Let R be an infinite set and suppose the n-element subsets of R are divided into k classes. Then there exists an infinite subset S of R such that all the n-element subsets of S belong to the same class.

If all we want to conclude is that R has some finite subset S with, say, s elements, where s is an integer $> n$, such that all the n-element subsets of S are in the same class, then R does not have to be infinite either. For instance, we just proved that if $n = 2, k = 3$ then there will be a subset S with $s = 3$ elements such that all pairs from this subset belong to the same class, provided R has at least 17 elements. For even slightly larger values of n, k or s, the number of elements in R must be enormous to ensure the existence of a subset S with the above property.

For a proof of Ramsey's Theorem the reader may consult H. J. Ryser: *Combinatorial Mathematics*, The Math. Assoc. of America, 1963.

1964/5 Solution (due to P. Herdeg). Choose any one of the 5 points. The remaining four determine $\binom{4}{2} = \frac{4 \cdot 3}{2} = 6$ lines, to each of which a perpendicular is to be drawn. Since there are 5 points in all, a total of $5 \times 6 = 30$ perpendiculars are drawn. The maximum number of intersection points is $\binom{30}{2} = \frac{30 \cdot 29}{2} = 435$. However, as we shall now show, not all of these intersection points are distinct.

(i) Consider one of the ten lines determined by the original 5 points. Two of the points lie on it, three do not. The three perpendiculars from these three points to the chosen line are parallel, hence do not intersect. Had they not been parallel, they could have intersected in three points. Since this occurs for each of 10 lines, $10 \times 3 = 30$ intersections are lost from the maximum number 435 of intersections counted above, so 405 are left.

(ii) The original 5 points determine $\binom{5}{3} = \frac{5 \cdot 4 \cdot 3}{1 \cdot 2 \cdot 3} = 10$ triangles. In each, the perpendiculars from the vertices to the opposite sides (the altitudes) are concurrent; they meet in one instead of 3 points. So two points of intersection are lost in each of 10 triangles. Thus there are 20 fewer intersections, leaving 385.

(iii) Each of the original 5 points has six perpendiculars passing through it—the six perpendiculars drawn from it to the six lines determined by the other 4 points. These six concurrent lines thus meet in one instead of $\binom{6}{2} = \dfrac{6 \cdot 5}{2} = 15$ points, so 14 points of intersection are lost for each of the five points, a total of $5 \cdot 14 = 70$. This leaves 315 possible intersections of the perpendiculars.

It is not hard to see that in general the lost intersections considered in (i), (ii) and (iii) fall into distinct categories and do not overlap. So we are justified in subtracting each set from our original count.

1964/6 First solution. In the tetrahedron $ABCD$ of Fig. 21, D_0 is the centroid of $\triangle ABC$, and AA_1, BB_1, CC_1 are lines parallel to DD_0 as specified by the problem. Now AA_1 intersects plane BCD at A_1. The plane determined by parallel lines AA_1 and DD_0 contains the median AM of $\triangle ABC$. Hence $\triangle MAA_1 \sim \triangle MD_0D$; and since $MA = 3MD_0$, we have $AA_1 = 3D_0D$. In the same manner, we can show that $BB_1 = 3D_0D$ and $CC_1 = 3D_0D$. It follows that AA_1B_1B, BB_1C_1C and CC_1A_1A are parallelograms, and that $\triangle A_1B_1C_1 \cong \triangle ABC$.

Now draw perpendiculars DP and A_1Q from D and A_1 to the plane ABC, and denote the lengths of these perpendiculars by h and h_1 respectively. Since $\triangle A_1AQ \sim \triangle DD_0P$, we have $h_1/h = AA_1/D_0D = 3$; thus $h_1 = 3h$.

The volume of a tetrahedron is equal to one-third the product of the area of its base and its altitude. So

$$\text{volume of } ABCD = \tfrac{1}{3}h(ABC)\ ,$$

and

$$\begin{aligned}\text{volume of } A_1B_1C_1D_0 &= \tfrac{1}{3}h_1(A_1B_1C_1) = \tfrac{1}{3}(3h)(ABC)\\ &= 3 \text{ times volume of } ABCD\ .\end{aligned}$$

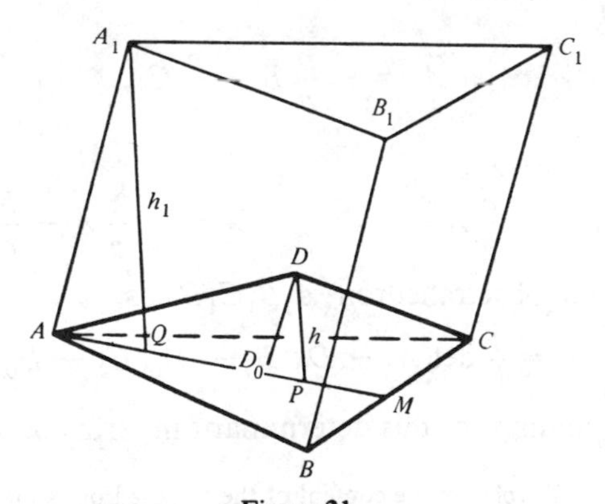

Figure 21

Note. This solves part of the given problem. The remaining part is not easily solved by the above method. Once one suspects that the result is true for any point in $\triangle ABC$ (this conjecture was the result of formulating and solving an analogous problem in two dimensions), one is encouraged to devise proofs in the more general situation. We shall consider the case where the point D_0 is any point in the plane of triangle ABC; and we shall find that the assertion of the problem remains true provided D_0 is not on the sides of $\triangle ABC$ or their extensions.

Our solution involves determinants of 3×3 matrices. The reader needs to know how to evaluate such determinants and how to multiply two matrices. We shall use this notation: Let U, V, W represent column vectors; then $\det(U, V, W)$ is the determinant of the matrix whose columns are U, V, W. Moreover, if U, V, W are vectors from the origin, then the volume of the tetrahedron whose vertices are the origin and the endpoints of U, V, W is given by

$$\text{(1)} \qquad \frac{1}{6}\det(U, V, W) = \text{volume of } OUVW\ .\dagger$$

Second solution. We place vertex D on the origin of a three-dimensional coordinate system. Then by (1) the volume of $ABCD$ is

$$v = \tfrac{1}{6}\det(A, B, C)\ .$$

If D_0 is in the plane of $A, B, C,$ it can be written in the form

$$D_0 = aA + bB + cC \qquad \text{with} \qquad a + b + c = 1\ ,$$

and if D_0 is not on the sides of $\triangle ABC$ nor on their extensions, then $abc \neq 0$. Now the line through A parallel to OD_0 may be represented parametrically by

$$\text{(2)} \quad A + tD_0 = A + t(aA + bB + cC) = (1 + at)A + btB + ctC\ ;$$

it meets the plane BCD when the coefficient of A in (2) vanishes, i.e. when $t = -\dfrac{1}{a}$. Therefore

$$A_1 = -\frac{b}{a}B - \frac{c}{a}C\ .$$

Similarly,

$$B_1 = -\frac{a}{b}A - \frac{c}{b}C, \qquad C_1 = -\frac{a}{c}A - \frac{b}{c}B\ .$$

By (1), the volume of tetrahedron $A_1B_1C_1D_0$ is

$$\text{(3)} \qquad v_1 = \frac{1}{6}\det(A_1 - D_0, B_1 - D_0, C_1 - D_0)\ .$$

We express the columns of this determinant in terms of A, B, C by means

†This yields a "signed" volume; the content of the tetrahedron is the absolute value of this expression.

of the above relations:

$$\begin{aligned} A_1 - D_0 &= -aA - \left(\frac{b}{a} + b\right)B - \left(\frac{c}{a} + c\right)C \\ B_1 - D_0 &= -\left(\frac{a}{b} + a\right)A - bB - \left(\frac{c}{b} + c\right)C \\ C_1 - D_0 &= -\left(\frac{a}{c} + a\right)A - \left(\frac{b}{c} + b\right)B - cC\,. \end{aligned}$$

Using the rules of matrix multiplication, one can verify that the matrix with columns $A_1 - D_0$, $B_1 - D_0$, $C_1 - D_0$ is the product of the matrix with columns A, B, C and the matrix $-M$, where

$$M = \begin{bmatrix} a & \frac{a}{b} + a & \frac{a}{c} + a \\ \frac{b}{a} + b & b & \frac{b}{c} + b \\ \frac{c}{a} + c & \frac{c}{b} + c & c \end{bmatrix}.$$

Since the determinant of the product of two matrices is the product of their determinants, we have

$$|\det(A_1 - D_0, B_1 - D_0, C_1 - D_0)| = |\det(A, B, C)\det(M)|\,.$$

We compute the determinant of M and find

$$\det M = 2 + a + b + c = 3\,.$$

It follows that

$$v_1 = 3v\,.$$

Seventh International Olympiad, 1965

1965/1. We observe that the inequality between the middle and right members holds for all x, since the absolute value of the difference of two positive numbers is at most equal to the larger. Moreover, the inequality between the left and middle members certainly holds for those x for which $\cos x \leqslant 0$, i.e. for $\pi/2 \leqslant x \leqslant 3\pi/2$. To find all other x for which it holds, we note that if $\cos x \geqslant 0$, the left inequality is equivalent to the one obtained by squaring both sides, i.e.

$$\begin{aligned} 4\cos^2 x &\leqslant 1 + \sin 2x - 2\sqrt{1 - \sin^2 2x} + 1 - \sin 2x \\ &= 2 - 2|\cos 2x|\,. \end{aligned}$$

Since $\cos^2 x = \dfrac{1 + \cos 2x}{2}$, this is equivalent to

$$2 + 2\cos 2x \leqslant 2 - 2|\cos 2x|\,,$$

or finally

$$|\cos 2x| \leqslant -\cos 2x \ .$$

Evidently this holds if and only if $\cos 2x \leqslant 0$, i.e. when x is in either of the intervals $[\pi/4, 3\pi/4]$ or $[5\pi/4, 7\pi/4]$. Combining these intervals with the earlier one $[\pi/2, 3\pi/2]$, we find that the complete solution set is $[\pi/4, 7\pi/4]$.

1965/2 First solution. Suppose the given system had a non-trivial solution; that is, at least one of the x's were not zero. We shall then arrive at a contradiction by showing that at least one of the given equations cannot be satisfied. We accomplish this by taking the x_i with largest absolute value in our alleged non-trivial solution and showing that the i-th equation cannot be satisfied because the term $a_{ii}x_i$ dominates in the sense that $a_{i1}x_1 + a_{i2}x_2 + a_{i3}x_3$ has the same sign as $a_{ii}x_i$, and hence is not zero.

For example, if $i = 2$, that is if

$$x_2 \neq 0 \ , |x_2| \geqslant |x_1| \ , |x_2| \geqslant |x_3| \ , \tag{1}$$

we show that the second equation cannot hold. Indeed, writing it in the form

$$a_{22}x_2 = -a_{21}x_1 - a_{23}x_3 \ ,$$

and using the triangle inequality, together with (a) and (b), we obtain

$$a_{22}|x_2| \leqslant -a_{21}|x_1| - a_{23}|x_3| \ .$$

Hence by (1),

$$a_{22}|x_2| \leqslant -a_{21}|x_2| - a_{23}|x_2| \ .$$

Dividing by $|x_2|$, we get $a_{22} \leqslant -a_{21} - a_{23}$, i.e., $a_{21} + a_{22} + a_{23} \leqslant 0$, contradicting (c).

Similarly, if x_1 has the largest absolute value, we can show that the first equation cannot hold, and if x_3 has the largest absolute value, the third equation cannot hold.

An $n \times n$ matrix (a_{ij}) satisfying the conditions (a) $a_{ii} > 0$, (b) $a_{ij} < 0$ for $i \neq j$, and (c) $\sum_{j=1}^{n} a_{ij} > 0$ for $i = 1, 2, \ldots, n$ is called diagonally dominant; such matrices come up in many applications.

Second solution. A system of linear homogeneous equations can have non-zero solutions only if the determinant of its coefficient matrix is zero. We shall show that the determinant

$$\Delta = \begin{vmatrix} a_{11} & a_{12} & a_{13} \\ a_{21} & a_{22} & a_{23} \\ a_{31} & a_{32} & a_{33} \end{vmatrix}$$

of the given system is positive by using some elementary column operations and the given conditions.

Replace the last column of $\triangle$ by the sum of all three columns, obtaining

$$(2) \qquad \triangle = \begin{vmatrix} a_{11} & a_{12} & s_1 \\ a_{21} & a_{22} & s_2 \\ a_{31} & a_{32} & s_3 \end{vmatrix}$$

where, by (c), $s_i = a_{i1} + a_{i2} + a_{i3} > 0$ for $i = 1, 2, 3$. We expand $\triangle$ as given in (2) with respect to its last column:

$$(3) \quad \triangle = s_1(a_{21}a_{32} - a_{22}a_{31}) + s_2(a_{12}a_{31} - a_{11}a_{32}) + s_3(a_{11}a_{22} - a_{12}a_{21}) \ .$$

It follows from conditions (a) and (b) that the first two terms in (3) are positive. Moreover, by (b) and (c)

$$a_{11} + a_{12} > a_{11} + a_{12} + a_{13} > 0 \ , \qquad \text{so} \qquad a_{11} > -a_{12} = |a_{12}| \ ,$$

and

$$a_{22} + a_{21} > a_{22} + a_{21} + a_{23} > 0 \ , \qquad \text{so} \qquad a_{22} > -a_{21} = |a_{21}| \ .$$

Hence $a_{11}a_{22} > a_{12}a_{21}$, so the third term of (3) is also positive. Thus $\triangle > 0$, and the system has only the trivial solution $x_1 = x_2 = x_3 = 0$.

1965/3 First solution. The volume of a polyhedron all of whose vertices lie in a pair of parallel planes is given by the formula†

$$(1) \qquad V = \frac{h}{6}(B_1 + B_2 + 4M) \ ,$$

where B_1 and B_2 are the areas of the bases, h is the altitude, and M is the area of the midsection. We shall apply this formula to each of the solids into which the plane ε divides the tetrahedron.

Since ε is parallel to AB and DC, it intersects the tetrahedron in a parallelogram $PQRS$ with $PQ \| SR \| CD$ and $QR \| PS \| AB$; see Figure 22a. Let m be the altitude of the polyhedron Π_1 with one base $PQRS$, the other base segment AB; and let n be the altitude of the polyhedron Π_2 with base $PQRS$ and the other base segment DC. Then $m + n = d$, the altitude of the given tetrahedron, and $m/n = k$, the given ratio. Denote the volumes of Π_1 and Π_2 by V_1 and V_2 respectively. The area of $PQRS$ is $(PQRS) = (PQ)(QR) \sin \omega$. Since $PQ/DC = m/d$ and $QR/AB = n/d$, we have

$$(2) \qquad PQ = \frac{bm}{d} \ , \qquad QR = \frac{an}{d} \ ,$$

†See the note following this solution for proofs of this prismatoid formula.

and

(3) $$(PQRS) = \frac{abmn}{d^2} \sin \omega .$$

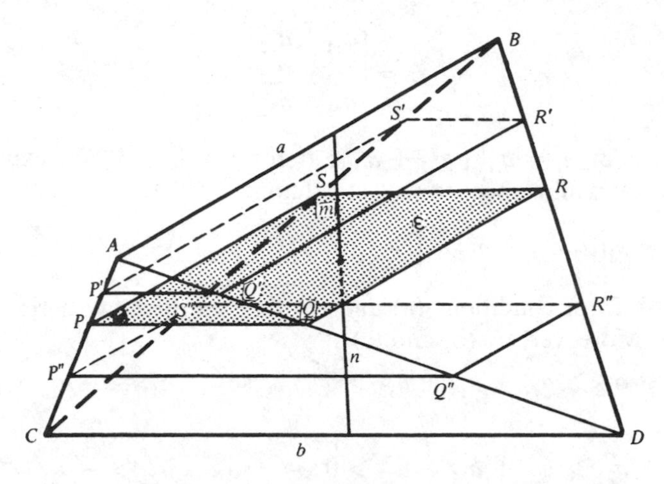

Figure 22a

We shall also need the areas of the midsections M_1 and M_2 of Π_1 and Π_2. These are parallelograms $P'Q'R'S'$ and $P''Q''R''S''$ with areas

$$M_m = \frac{1}{2}(PQ)\left(\frac{QR + a}{2}\right) \sin \omega , \qquad M_n = \frac{1}{2}(QR)\left(\frac{PQ + b}{2}\right) \sin \omega .$$

Substituting for PQ and QR from (2), we obtain

(4) $$4M_m = \frac{bm}{d}\left(\frac{an}{d} + a\right) \sin \omega = \frac{abm}{d^2}(n + d) \sin \omega ,$$

(5) $$4M_n = \frac{an}{d}\left(\frac{bm}{d} + b\right) \sin \omega = \frac{abn}{d^2}(m + d) \sin \omega .$$

For the volumes, we obtain, using (3), (4) and (5)

$$V_m = \frac{m}{6}\left[(PQRS) + 4M_m\right] = \frac{m^2ab}{6d^2}(2n + d) \sin \omega$$

$$V_n = \frac{n}{6}\left[(PQRS) + 4M_n\right] = \frac{n^2ab}{6d^2}(2m + d) \sin \omega ,$$

and their ratio is

(6) $$\frac{V_m}{V_n} = \frac{m^2}{n^2}\,\frac{2n + d}{2m + d} .$$

Since $m/n = k$ and $m + n = d$, we have

$$\frac{n}{d} = \frac{1}{k+1} \quad \text{and} \quad \frac{m}{d} = \frac{k}{k+1}.$$

We can therefore rewrite (6), the desired ratio, in the form

$$\frac{V_m}{V_n} = k^2 \frac{(2n/d) + 1}{(2m/d) + 1} = k^2 \frac{2 + k + 1}{2k + k + 1} = k^2 \frac{k+3}{3k+1}.$$

Note. We now present a geometric and then an analytic proof of the prismatoid formula (1).

A prismatoid is a polyhedron all of whose vertices lie in two parallel planes, β and γ. The vertices $B_1, B_2, \ldots, B_n$ (listed in cyclic order) in plane β are joined by edges B_iC_i to the vertices $C_1, C_2, \ldots, C_n$ in the plane γ. Here we have used the convention that several consecutive B's or C's could coincide. The lateral faces are generally triangles, occasionally quadrangles, see Figure 22b.

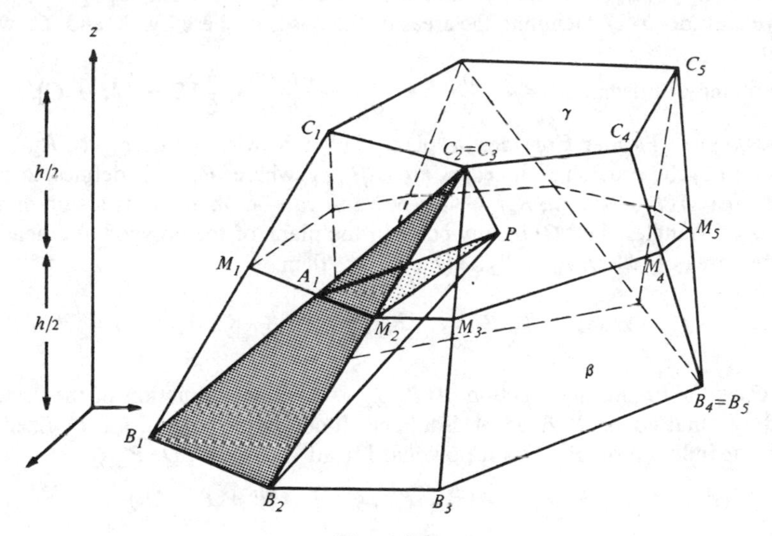

Figure 22b

GEOMETRIC PROOF: Divide each lateral quadrangle into two triangles by a diagonal (as in $B_1B_2C_2C_1$ with diagonal B_1C_2). Denote by M the polygonal midsection $M_1M_2 \ldots M_n$ of the prismatoid. From a point P in M, draw lines to all vertices M_i of M, to all points A_i where diagonals of faces intersect M, and to all the vertices B_i and C_i of the prismatoid. We shall determine the volume W_1 of the pyramid $PB_1B_2C_2$.

Since A_1 and M_2 are midpoints of B_1C_2 and B_2C_2, respectively, we have $(B_1B_2C_2) = 4(A_1M_2C_2)$; hence $W_1 = 4$ Vol $(PA_1M_2C_2)$, where $PA_1M_2C_2$ can also be viewed as a pyramid with base PA_1M_2 and altitude $h/2$ from C_2, so that

$$W_1 = 4 \cdot \tfrac{1}{3}(PA_1M_2)\frac{h}{2} = 4 \times \frac{h}{6} m_1 , \tag{7}$$

where $m_1 = (PA_1M_2) =$ area of $\triangle PA_1M_2$.

By the same method, we find the volumes $W_2, W_3, \ldots,$ of pyramids $PB_2B_3C_2$, $PB_3B_4C_3, \ldots,$ until we have expressed the volume W_i of each pyramid with vertex P and base on a lateral triangle as

$$W_i = 4 \times \frac{h}{6} m_i \tag{7$_i$}$$

where m_i is the area of triangle PM_iM_j (or PM_iA_j) in M. All these triangles appearing on the right in $(7)_i$ fill up M; so adding all these volumes W_i, we find that

$$\sum W_i = 4 \cdot \frac{h}{6} \sum m_i = 4 \cdot \frac{h}{6} M .$$

To fill up the prismatoid, we must include also the pyramid with vertex P and base $B_1B_2 \ldots B_n$ and the pyramid with vertex P and base $C_1C_2 \ldots C_n$. Both have altitude $h/2$. Denoting the areas of the bases of these by B and C, we find that

$$\text{Volume of prismatoid} = 4 \cdot \frac{h}{6} M + \frac{h}{6} B + \frac{h}{6} C = \frac{h}{6}[B + 4M + C].$$

Analytic Proof: Consider a plane polygon with vertices $R_1, R_2, \ldots, R_n$ listed in cyclic order; i.e. the edges are R_iR_{i+1} where R_{n+1} is defined to be R_1. The area $A(R_1, R_2, \ldots, R_n)$ of this polygon can be obtained as a sum of signed areas of triangles: Let O be any point in the plane of the polygon and denote the signed area of $\triangle OR_jR_{j+1}$ by $A(R_j, R_{j+1})$; then

$$A(R_1, R_2, \ldots, R_n) = \sum_1^n A(R_j, R_{j+1}) . \tag{8}$$

If O is the origin, the function $A(P, Q)$ is a bilinear function of the vectors P and Q; that is, for P fixed A is a linear function of Q, and for Q fixed A is a linear function of P. This means that for all vectors P, Q, P', Q'

$$\begin{aligned} A(P + P', Q + Q') &= A(P + P', Q) + A(P + P', Q') \\ &= A(P, Q) + A(P', Q) + A(P, Q') + A(P', Q') , \end{aligned} \tag{9}$$

and for all numbers p and q,

$$A(pP, qQ) = pqA(P, Q) . \tag{10}$$

Now we place the prismatoid so that one of its parallel bases lies in the plane $z = 0$, the other in the plane $z = h$, see Figure 22b. The edge B_iC_i intersects the plane at height z in the point

$$R_i(z) = \frac{z}{h} C_i + \frac{h - z}{h} B_i .$$

[In Figure 22b, $R_i(0) = B_i$, $R_i(h/2) = M_i$, $R_i(h) = C_i$.] The cross section of the prismatoid at height z is the polygon with vertices $R_1(z), R_2(z), \ldots, R_n(z)$

whose area, by (8) and properties (9) and (10) is

$$A(z) = A(R_1(z), R_2(z), \dots, R_n(z)) = \sum_{i=1}^{n} A(R_i(z), R_{i+1}(z))$$

$$= \sum_{i=1}^{n} A\left(\frac{z}{h} C_i + \frac{h-z}{h} B_i,\quad \frac{z}{h} C_{i+1} + \frac{h-z}{h} B_{i+1}\right)$$

$$= \frac{1}{h^2} \sum_{i=1}^{n} \{z^2 A(C_i, C_{i+1}) + z(h-z)[A(C_i, B_{i+1}) + A(B_i, C_{i+1})]$$

$$+ (h-z)^2 A(B_i, B_{i+1})\},$$

clearly a quadratic function of z.

The volume V of a solid can be calculated by the integral

$$\int_0^h A(z)dz \, . \tag{11}$$

For a quadratic integrand, Simpson's rule gives the exact value of the integral. Applied to the integral (11), Simpson's rule gives

$$V = \int_0^h A(z)dz = h\left(\frac{1}{6} A(0) + \frac{4}{6} A\left(\frac{h}{2}\right) + \frac{1}{6} A(h)\right),$$

and this is the prismatoid formula with $A(0)$, $A(h)$ denoting the areas of the bases and $A\left(\frac{h}{2}\right)$ that of the midsection.

Second solution. An affine transformation† of space sends $ABCD$ into another tetrahedron, and ε into another plane. However it preserves both the ratio k and the ratio of the volumes of the two pieces into which ε divides $ABCD$. Since any tetrahedron can be mapped onto any other by an affine transformation, we may therefore suppose from the outset that $A = (1, 0, 0)$, $B = (0, 1, 0)$, $C = (0, 0, 1)$, and $D = (0, 0, 0)$. This gives the situation shown in Figure 22c. The plane ε intersects the tetrahedron in a rectangle $PQRS$. If we set $DP = t$, then $PB = 1 - t$, and hence $k = t/(1-t)$. To find the volume of the portion of $ABCD$ "behind" ε, we break it into the prism bounded by DPS and EQR, and the tetrahedron $CEQR$. The base DPS of the prism has area $t^2/2$, and the height of the prism is $PQ = PB = 1 - t$. Hence the prism has volume $t^2(1-t)/2$. The base EQR of the tetrahedron $CEQR$ has area $t^2/2$, and the altitude $CE = CD - ED = 1 - (1-t) = t$. Hence the volume of $CEQR$ is

$$\frac{1}{3}\left(\frac{1}{2}t^2\right)t = \frac{t^3}{6} \, .$$

†For a definition of affine transformations and a list of their properties, see e.g. *Geometric Transformations III* by I. M. Yaglom, NML vol. 24, 1973, p. 9 ff.

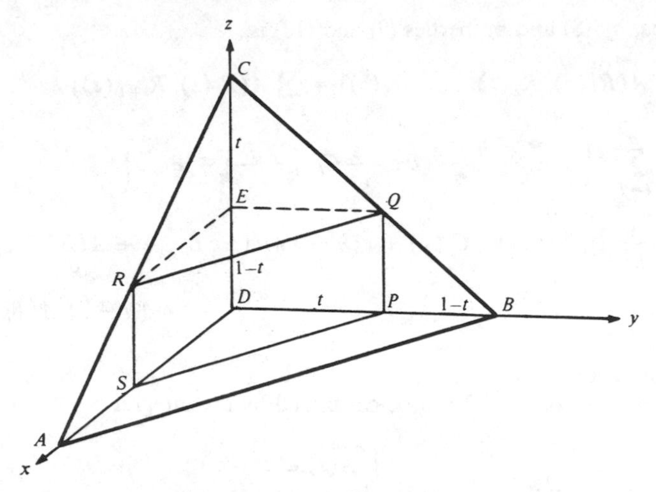

Figure 22c

It follows that the portion of $ABCD$ behind ε has volume

$$\frac{1}{2}t^2(1-t)+\frac{t^3}{6}=\frac{t^2(3-2t)}{6}\ .$$

Since the volume of $ABCD$ is $1/6$, the portion in front of ε has volume

$$\frac{1}{6}-\frac{t^2(3-2t)}{6}=\frac{(1-t)^2(1+2t)}{6}\ .$$

Thus the ratio of volumes of the two portions is $t^2(3-2t)/(1-t^2)(1+2t)$. Using the fact that $k=t/(1-t)$ and hence $t=k/(1+k)$, we find that this ratio is equal to

$$k^2\frac{3-[2k/(1+k)]}{1+[2k/(1+k)]}=k^2\frac{k+3}{3k+1}\ .$$

1965/4. We seek solutions to the system

$$\begin{aligned}x_1+x_2x_3x_4&=2\\x_2+x_1x_3x_4&=2\\x_3+x_1x_2x_4&=2\\x_4+x_1x_2x_3&=2\ ,\end{aligned}$$

which is symmetric in the four numbers x_1, x_2, x_3, x_4. Denote the product of all four numbers by p:

$$x_1x_2x_3x_4=p\ .$$

No x_i is zero; for if, say, $x_1=0$, then $x_2x_3x_4=2$ by the first equation, while $x_2 = x_3 = x_4 = 2$ by the other three, a contradiction. So we can write each equation in the form

$$x_i + \frac{p}{x_i} = 2\,, \qquad i = 1, 2, 3, 4\,,$$

or

$$(1) \qquad x_i^2 - 2x_i + p = 0\,, \qquad p \neq 0\,.$$

This quadratic equation for x_i has at most two distinct solutions,

$$(2) \qquad x_i = 1 \pm \sqrt{1-p}\;;$$

in order for them to be real, we must have $p \leqslant 1$. This leads to the following possibilities:

(a) The roots of (1) are equal, that is, $p = 1$. Then by (2), $x_1 = x_2 = x_3 = x_4 = 1$.

(b$_1$) The roots of (1) are distinct, two of the four numbers x_i are $1 + \sqrt{1-p}$, and the other two are $1 - \sqrt{1-p}$. Then

$$p = x_1x_2x_3x_4 = \left(1 + \sqrt{1-p}\,\right)^2\left(1 - \sqrt{1-p}\,\right)^2 = \left[1 - (1-p)\right]^2 = p^2\,.$$

Hence $p = 1$ which implies that the roots of (1) are equal, a contradiction.

(b$_2$) The two roots of (1) are distinct, one of them is the value of three of our numbers x_i, and the other is the value of the fourth. Then either

$$p = x_1x_2x_3x_4 = (1 + \sqrt{1-p}\,)^3(1 - \sqrt{1-p}\,) = p(1 + \sqrt{1-p}\,)^2$$

or

$$p = (1 + \sqrt{1-p}\,)(1 - \sqrt{1-p}\,)^3 = p(1 - \sqrt{1-p}\,)^2.$$

In the first case, we again get $p = 1$, which contradicts the fact that (1) has distinct roots. But in the second case, we have

$$\left(1 - \sqrt{1-p}\,\right)^2 = 1\,,$$

and this is satisfied also by $p = -3$. So by (2), one of the numbers x_i is equal to 3, and the other three are equal to -1. Thus, the set of all solutions (x_1, x_2, x_3, x_4) is

$$\{(1, 1, 1, 1), (3, -1, -1, -1), (-1, 3, -1, -1), (-1, -1, 3, -1), (-1, -1, -1, 3)\}\,.$$

1965/5. (a) We take O as the origin and denote by A, B, H, M, P, Q the vectors $\overrightarrow{OA}$, $\overrightarrow{OB}$, $\cdots$. In Figure 23a, $MP \| QH$, since both are perpendicular to OA. Similarly $MQ \parallel PH$, and so $MPHQ$ is a parallelogram. Hence $H - P = Q - M$, or in other words

(1) $$H = P + Q - M\ .$$

Now if $AM/AB = t$, we have

(2) $$M = A + t(B - A) = (1 - t)A + tB\ .\dagger$$

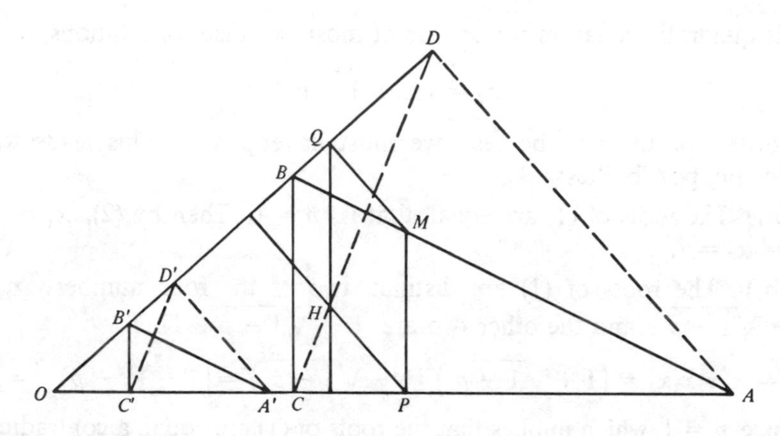

Figure 23

Let C be the foot of the altitude from B to OA, and D the foot of the altitude from A to OB (see Fig. 23a). We assert that

(3) $$P = (1 - t)A + tC, \qquad \text{and}$$

(4) $$Q = tB + (1 - t)D\ .$$

To prove (3), let us tentatively write

$$P_1 = (1 - t)A + tC\ ;$$

clearly P_1 is on OA. From (2) we have

$$M - P_1 = t(B - C)\ .$$

Thus the segment P_1M is parallel to CB, and is therefore perpendicular to OA. This proves that $P_1 = P$, as claimed. The proof of (4) is completely analogous.

Substituting (2), (3) and (4) into (1), we obtain

$$H = (1 - t)A + tC + tB + (1 - t)D - (1 - t)A - tB\ ,$$

†See the solution of 1960/5, p. 31 for more on this parametric vector representation of a segment AB.

or

$$(5) \qquad H = tC + (1 - t)D\ .$$

As t runs from 0 to 1, M runs from A to B, and by (5), H runs from D to C. Thus the locus of H is the line segment DC.

(b) We can view $\triangle OAB$ as the union of line segments $A'B'$ parallel to AB, as indicated in Figure 23. For a typical such segment $A'B'$, let C' be the foot of the altitude from B' to OA, and D' the foot of the altitude from A' to OB. By part (a), as M runs along $A'B'$, H runs along $D'C'$. Therefore as M runs through the whole triangle OAB, H sweeps out the triangle ODC.

1965/6. We shall prove the result by induction on n, by means of the following lemma.

LEMMA. *If more than two diameters issue from one of the given points, then there is another point from which only one diameter issues.*

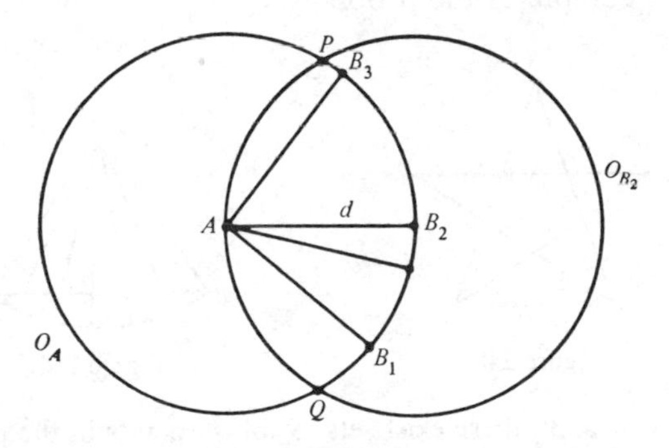

Figure 24a

PROOF: Let A be an endpoint of three (or more) diameters, see Figure 24a. The other endpoints of these diameters lie on a circle O_A with center A and radius d. Moreover, they all lie on an arc of radian measure $\leqslant \pi/3$, since otherwise the pair furthest apart would be at a distance $> d$ from each other. Denote the other endpoints of three diameters from A by B_1, B_2, B_3 where B_2 lies between B_1 and B_3 on this arc. With B_2 as center, draw a circle O_{B_2} with radius d, and denote the intersections of O_{B_2} and O_A by P and Q. We claim that no point of the given set, except A, lies on the circle O_{B_2}. For all points of the major arc PQ (except P and Q) are farther than d away from A, all points on arc PA (including P but not A) are farther than d away from B_1, and all points on arc QA (including Q but not A) are farther than d away from B_3. It follows that B_2A is the only diameter issuing from B_2. Thus if $k > 2$ diameters issue from A, there is at least one point from which only one diameter issues.

INDUCTION PROOF: For a set of three points, there are obviously at most three diameters. So the assertion of the problem holds for $n = 3$. Suppose it holds for sets of n points with $n = 3, 4, \ldots m$. We shall show that it then holds for sets of $m + 1$ points.

Consider a set S of $m + 1$ points. We distinguish two cases:

(a) At most two diameters issue from each of the $m + 1$ points. Since each diameter has two endpoints, there are at most $2(m + 1)/2 = m + 1$ diameters, so the assertion of the problem holds for S.

(b) There is a point A of S from which more than two diameters issue. Then, by the lemma proved above, there is another point B of S from which only one diameter issues. Now consider the set $S - B$ of m points remaining when B is deleted from S. By the induction hypothesis, $S - B$ has at most m diameters. When B is added to $S - B$, the resulting set S gains exactly one diameter. Hence S has at most $m + 1$ diameters. This completes the proof.

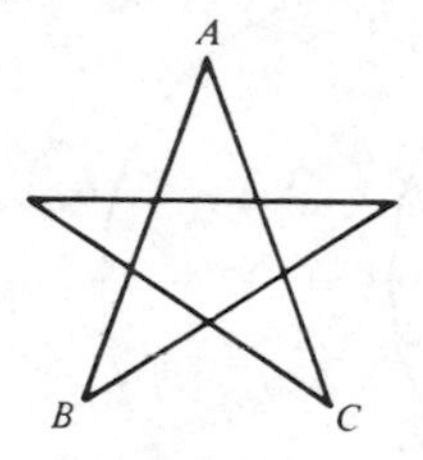

Figure 24b

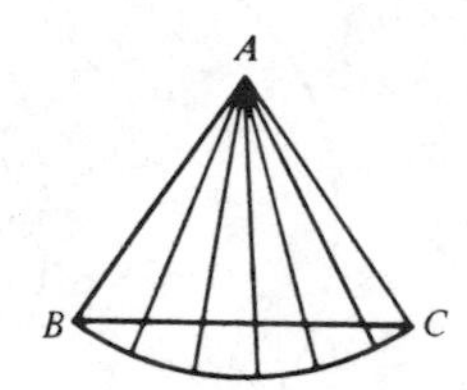

Figure 24c

Remark. For any $n \geqslant 3$, there exist sets S of n points in the plane with exactly n diameters. If n is odd, the set S of vertices of a regular n-gon has this property. (see Figure 24b, where $n = 5$.) To get an example which works for all $n \geqslant 3$, consider Figure 24c. In this figure A, B, C are the vertices of an equilateral triangle. The remaining $n - 3$ points are chosen on the circular arc BC with center A. Still further examples can be obtained by combining the ideas embodied in the two figures. Thus in Figure 24b we may add points on the arc BC with center A.

We note incidentally that Figures 24b and 24c illustrate the two cases (a) and (b) occurring in our induction proof.

Eighth International Olympiad, 1966

1966/1. The three circular disks marked A, B, C in Fig. 25 represent the sets of contestants who solved problems A, B, C respectively. The lower case letters denote the numbers of contestants in the sets labelled;

thus p, labelling the region common to B and C but outside A, is the number of students who solved problems B and C but not problem A. We wish to find the number y of students who solved only B. We first translate the given information into equations involving the above unknowns. Since 25 students solved at least one problem, we add the numbers of students who solved precisely one problem, precisely two, and all three to obtain

$$p + q + r + s + x + y + z = 25 . \tag{1}$$

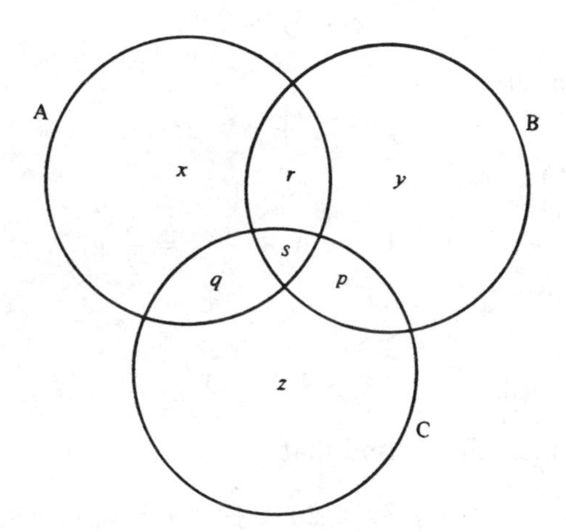

Figure 25

Among the $p + y + z$ students who did not solve A, $p + y$ solved B and $p + z$ solved C. Hence by hypothesis $p + y = 2(p + z)$, or

$$-p + y - 2z = 0 . \tag{2}$$

The number of students who solved only A is x, and the number who solved A and at least one other problem is $q + r + s$. According to the given information,

$$x = q + r + s + 1 . \tag{3}$$

Finally, $x + y + z$ students solved just one problem, and of these, $y + z$ did not solve A. We are told that $\frac{1}{2}(x + y + z) = y + z$, or

$$x = y + z . \tag{4}$$

Our four equations involve seven unknowns, but fortunately q, r and s appear only in the combination $q + r + s$. Hence we set $q + r + s = t$, and thereby obtain the system

$$\begin{array}{ll}(1') & p+t+x+y+z = 25\\ (2') & -p \quad +y-2z= 0\\ (3') & -t+x \quad = 1\\ (4') & -x+y+z = 0\end{array}$$

of four equations in five unknowns, each of which is a non-negative integer. Adding all four equations, we get

$$(5) \qquad x + 3y = 26 .$$

From (2′) we see that

$$y = 2z + p \geqslant 2z ,$$

and therefore (4′) yields

$$x = y + z \leqslant y + \frac{1}{2} y = \frac{3}{2} y .$$

Thus

$$(6) \qquad y \leqslant x \leqslant \frac{3}{2} y .$$

Substituting (6) into (5), we find that

$$y + 3y \leqslant 26 \leqslant \frac{3}{2} y + 3y .$$

The first inequality gives $4y \leqslant 26$, so $y \leqslant 13/2 < 7$. The second inequality gives $26 \leqslant 9y/2$, so $y \geqslant 52/9 > 5$. Hence $y = 6$.

1966/2 First solution. In the given equation, we replace $\tan x$ by $\sin x/\cos x$; then multiply both sides by $\cos\alpha \cos\beta \cos\frac{\gamma}{2}$. The result is

$$(a + b) \cos\alpha \cos\beta \cos\frac{\gamma}{2} = a \sin\alpha \cos\beta \sin\frac{\gamma}{2} + b \sin\beta \cos\alpha \sin\frac{\gamma}{2} ,$$

which is equivalent to

$$a \cos\beta\left(\cos\alpha \cos\frac{\gamma}{2} - \sin\alpha \sin\frac{\gamma}{2}\right) + b \cos\alpha\left(\cos\beta \cos\frac{\gamma}{2} - \sin\beta \sin\frac{\gamma}{2}\right) = 0 .$$

This, in turn, is equivalent to

$$a \cos\beta \cos\left(\alpha + \frac{\gamma}{2}\right) + b \cos\alpha \cos\left(\beta + \frac{\gamma}{2}\right) = 0 .$$

Since $\alpha + \frac{\gamma}{2} + \beta + \frac{\gamma}{2} = \alpha + \beta + \gamma = \pi$, we have

$$\cos\left(\beta + \frac{\gamma}{2}\right) = -\cos\left(\alpha + \frac{\gamma}{2}\right),$$

and therefore

$$(a \cos \beta - b \cos \alpha) \cos\left(\alpha + \frac{\gamma}{2}\right) = 0 .$$

If $\cos(\alpha + \gamma/2) = 0$, then $\alpha + \gamma/2 = \pi/2$, so $\beta + \gamma/2 = \pi/2$, and we conclude that $\alpha = \beta$. If $a \cos \beta - b \cos \alpha = 0$, then we use the law of sines $a \sin \beta = b \sin \alpha$ and divide it by $a \cos \beta = b \cos \alpha$ to deduce $\tan \alpha = \tan \beta$. From this it follows that $\alpha = \beta$, and the triangle is isosceles.

Second solution. By the law of sines,

$$\frac{a}{\sin \alpha} = \frac{b}{\sin \beta} = 2R .$$

Then $a = 2R \sin \alpha$, $b = 2R \sin \beta$. Substituting these expressions for a and b into the given relation and dividing both sides by $2R$ gives

$$\sin \alpha + \sin \beta = \tan \frac{\gamma}{2}\left(\frac{\sin^2\alpha}{\cos \alpha} + \frac{\sin^2\beta}{\cos \beta}\right). \tag{1}$$

Since $\alpha + \beta + \gamma = \pi$, we have $\gamma/2 = \pi/2 - (\alpha + \beta)/2$, and

$$\tan \frac{\gamma}{2} = \tan\left(\frac{\pi}{2} - \frac{\alpha + \beta}{2}\right) = \cot \frac{\alpha + \beta}{2} . \tag{2}$$

Next we write α and β as sums

$$\alpha = \frac{\alpha + \beta}{2} + \frac{\alpha - \beta}{2}, \qquad \beta = \frac{\alpha + \beta}{2} - \frac{\alpha - \beta}{2}$$

and apply the addition formulas for sin and cos:

$$\sin \alpha = \sin \frac{\alpha + \beta}{2} \cos \frac{\alpha - \beta}{2} + \cos \frac{\alpha + \beta}{2} \sin \frac{\alpha - \beta}{2},$$

$$\sin \beta = \sin \frac{\alpha + \beta}{2} \cos \frac{\alpha - \beta}{2} - \cos \frac{\alpha + \beta}{2} \sin \frac{\alpha - \beta}{2},$$

and

$$\cos \alpha = \cos \frac{\alpha + \beta}{2} \cos \frac{\alpha - \beta}{2} - \sin \frac{\alpha + \beta}{2} \sin \frac{\alpha - \beta}{2},$$

$$\cos \beta = \cos \frac{\alpha + \beta}{2} \cos \frac{\alpha - \beta}{2} + \sin \frac{\alpha + \beta}{2} \sin \frac{\alpha - \beta}{2} .$$

The sums of the first two and the last two are

$$\sin\alpha + \sin\beta = 2\sin\frac{\alpha+\beta}{2}\cos\frac{\alpha-\beta}{2}$$

and

$$\cos\alpha + \cos\beta = 2\cos\frac{\alpha+\beta}{2}\cos\frac{\alpha-\beta}{2}$$

respectively, and the ratio of the second sum to the first is

$$\frac{\cos\alpha+\cos\beta}{\sin\alpha+\sin\beta} = \cot\frac{\alpha+\beta}{2} = \tan\frac{\gamma}{2}\ .$$

We substitute this expression for $\tan\frac{\gamma}{2}$ into (1) and obtain, after multiplying both sides by $\sin\alpha + \sin\beta$,

$$(3) \qquad (\sin\alpha+\sin\beta)^2 = (\cos\alpha+\cos\beta)\left(\frac{\sin^2\alpha}{\cos\alpha} + \frac{\sin^2\beta}{\cos\beta}\right) .$$

Equation (3) is easily brought into the form

$$2\sin\alpha\sin\beta\cos\alpha\cos\beta = \cos^2\alpha\sin^2\beta + \cos^2\beta\sin^2\alpha\ ,$$

which is equivalent to

$$(\sin\alpha\cos\beta - \cos\alpha\sin\beta)^2 = 0\ .$$

The expression on the left is $\sin^2(\alpha-\beta)$, and this vanishes only if angles α and β are equal. We conclude that $\alpha = \beta$, and therefore $\triangle ABC$ is isosceles with $a = b$.

Third solution. Since $\gamma = \pi - \alpha - \beta$, we have

$$\tan\frac{\gamma}{2} = \tan\left(\frac{\pi}{2} - \frac{\alpha+\beta}{2}\right) = \cot\frac{\alpha+\beta}{2}\ .$$

Hence the given equation can be written in the form

$$a + b = \cot\frac{\alpha+\beta}{2}(a\tan\alpha + b\tan\beta)$$

or, equivalently,

$$(4) \qquad (a+b)\tan\frac{\alpha+\beta}{2} = a\tan\alpha + b\tan\beta\ .$$

We can suppose without loss of generality that $\alpha \leqslant \beta$, and therefore $a \leqslant b$. We first note that α and β must be acute angles; for, if β were obtuse, the right side of (4) would be negative (since then $a < b$ and $\tan\alpha < \tan(\pi-\beta) = |\tan\beta|$), while the left side of (4) is positive.

Now for acute angles, we have the inequality

$$(5) \qquad \tan\frac{\alpha+\beta}{2} \leqslant \frac{\tan\alpha+\tan\beta}{2},$$

where equality holds if and only if $\alpha = \beta$. We shall prove inequality (5) in the note below.

If $\alpha < \beta$, substitution of (5) into (4) yields

$$\frac{a+b}{2}(\tan\alpha+\tan\beta) > a\tan\alpha + b\tan\beta$$

or

$$\frac{b-a}{2}\tan\alpha > \frac{b-a}{2}\tan\beta$$

or, finally, $\tan\alpha > \tan\beta$. Since $\tan x$ is an increasing function for $0 < x < \pi/2$, this contradicts $\alpha < \beta$. We conclude that $\alpha = \beta$, i.e. the triangle is isosceles.

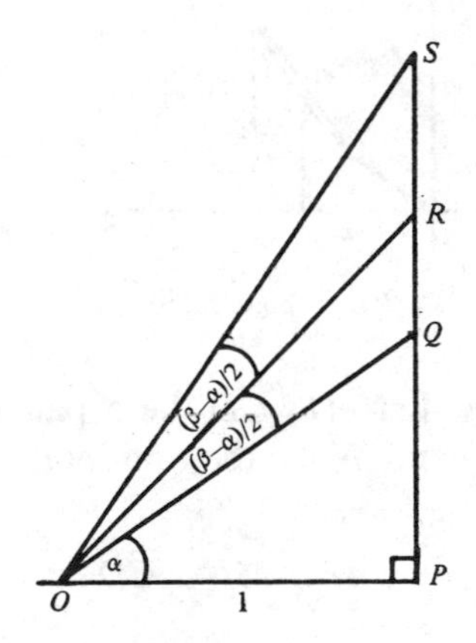

Figure 26a

Note. We give two proofs of inequality (5). The first is geometric and can be seen from Fig. 26a, where $OP = 1$, $\alpha \leqslant \beta$, and

$$(6) \quad PQ = \tan\alpha, \quad PS = \tan\beta, \quad PR = \tan\left(\alpha + \frac{\beta-\alpha}{2}\right) = \tan\frac{\alpha+\beta}{2}.$$

Clearly $QR \leqslant RS$, i.e. $PR - PQ \leqslant PS - PR$. Hence

$$PR \leqslant \frac{PQ+PS}{2}$$

By (6), the last inequality is equivalent to (5).

The second proof consists of observing that $y = \tan x$ is a *convex function* in the interval $(0, \pi/2)$. This means that for any two points $\alpha, \beta, \alpha < \beta$ in this interval, the graph of $y = \tan x$ $(\alpha \leqslant x \leqslant \beta)$ lies below the line segment connecting the points $(\alpha, \tan \alpha)$ and $(\beta, \tan \beta)$, see Fig. 26b. In particular, the point

$$L = \left(\frac{\alpha + \beta}{2}, \tan \frac{\alpha + \beta}{2} \right)$$

lies below the midpoint

$$M = \left(\frac{\alpha + \beta}{2}, \frac{\tan \alpha + \tan \beta}{2} \right)$$

of that segment. The analytic criterion for convexity of a function is that its second derivative is positive. Indeed, $y' = \sec^2 x$ and $y'' = \sec^2 x \tan x > 0$ for $0 < x < \pi/2$.

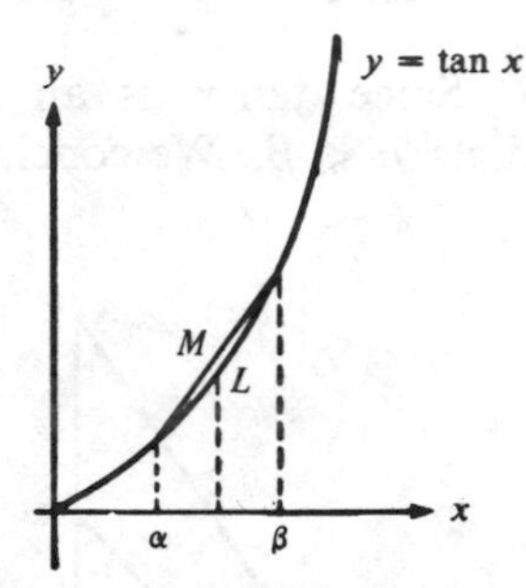

Figure 26b

1966/3 First solution. Let AB, CD be a pair of opposite edges of the tetrahedron, and let M and N be their midpoints; see Fig. 27a. We claim that

$$(1) \qquad AB \perp MN \perp CD.$$

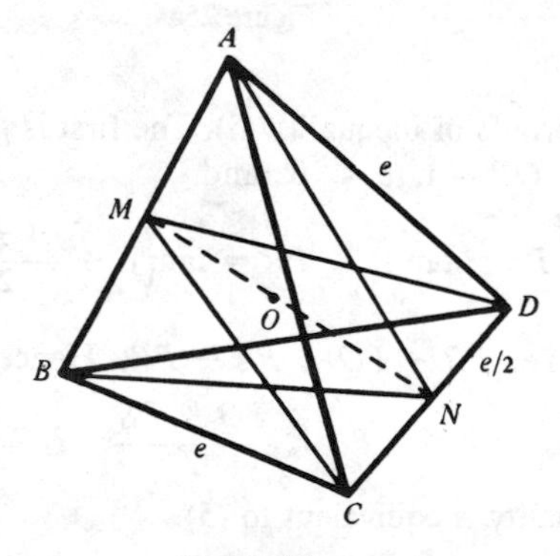

Figure 27a

To see this, denote the edge-length of the tetrahedron by e, and note that

$AN = e\sqrt{3}/2 = BN$ and $CM = e\sqrt{3}/2 = DM$. Thus NM is a median of isosceles $\triangle ABN$, hence perpendicular to its base AB; it is also a median of isosceles $\triangle CMD$ and perpendicular to its base CD. Since MN is perpendicular to both AB and CD, its length is the distance between these two skew lines. We denote this length by p.

We choose any point P, form the sum

$$s = PA + PB + PC + PD , \tag{2}$$

and show that this sum is smallest when P is the midpoint of MN, i.e. the center of the circumscribed sphere of our regular tetrahedron.

Let h_1 and h_2 be the altitudes from P of triangles APB and CPD, respectively. Since the distance of any point on AB from any point on CD is at least equal to p, we have $h_1 + h_2 \geqslant p$. Now draw two isosceles triangles $A'P'B'$ and $C'P'D'$ in the same plane with bases $A'B' = C'D' = e$ and altitudes from P' of lengths h_1 and h_2. Place them so that $A'B' \| C'D'$, and their common vertex P' lies in the strip between these bases, see Fig. 27b. Triangle $A'P'B'$ has the same base and altitude as $\triangle APB$, and similarly for $\triangle C'P'D'$ and $\triangle CPD$. We now make use of the fact that, among all triangles with given base and altitude, the isosceles triangle has the smallest perimeter. (For a proof, see N. Kazarinoff, *Geometric Inequalities*, vol. 4 of this NML series, Theorem 10A, or contemplate Figure 27c.) This implies that

$$s \geqslant P'A' + P'B' + P'C' + P'D' . \tag{3}$$

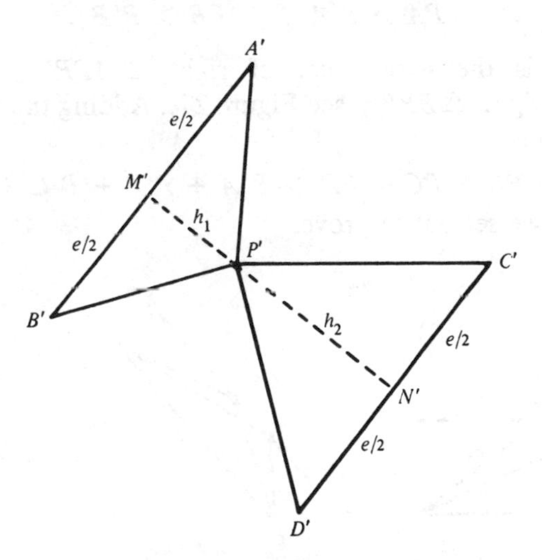

Figure 27b

Moreover, since $A'P' + P'D' \geqslant A'D'$ and $B'P' + P'C' \geqslant B'C'$ (see Fig. 27b),

(4) $$s \geqslant A'D' + B'C' ,$$

$A'D'$ and $B'C'$ being diagonals of a rectangle with sides e and $h_1 + h_2 \geqslant p$. If P is the midpoint of MN, then APB and CPD are isosceles triangles with height $p/2$; in this case our rectangle has dimensions $e \times p$, and the sum $2(e^2 + p^2)^{1/2}$ of its diagonals is exactly s. If P is any other point, at least one of the inequalities (3), (4) is strict. So $s > 2(e^2 + p^2)^{1/2}$ unless P is the midpoint of MN.

Second solution. The center O of the circumsphere of a regular tetrahedron is the point of intersection of six planes, each through an edge and the midpoint of the opposite edge of the tetrahedron. We shall show that, if a point P fails to be on one of these planes (i.e. if it is not O), then the sum (2) is not minimal. From this we conclude that the point that minimizes s lies on all these planes, hence must be the point O.†

Suppose P does not lie on plane ABN (Fig. 27a), where N is the midpoint of CD. Let l be a line through P parallel to CD, hence perpendicular to plane ABN, and let P' be the point where l intersects ABN. Then

(5) $$PC + PD > P'C + P'D ;$$

indeed triangles CPD and $CP'D$ have the same base and altitude, and since the latter is isosceles, it has the smaller perimeter (see first solution and Figures 27c and d.) Moreover,

(6) $$PA > P'A , \qquad PB > P'B ,$$

because PA is the hypotenuse of right $\triangle APP'$, and PB is the hypotenuse of right $\triangle BPP'$; see Figure 27e. Adding the three inequalities (5) and (6) yields

$$PA + PB + PC + PD > P'A + P'B + P'C + P'D ,$$

which is what we set out to prove.

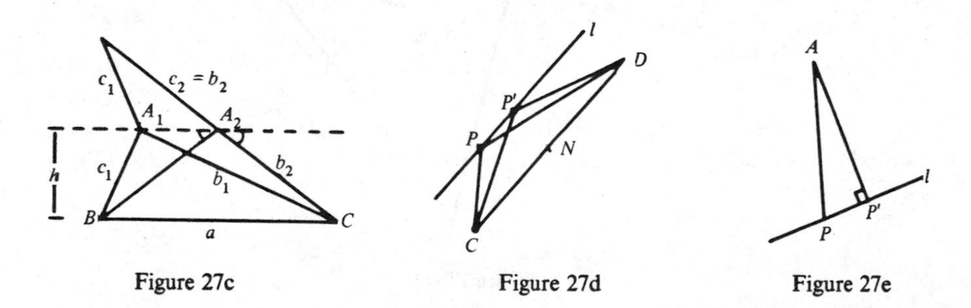

Figure 27c Figure 27d Figure 27e

†Strictly speaking, this argument shows only that if there is a minimum, it is achieved when P is the center O of the circumsphere.

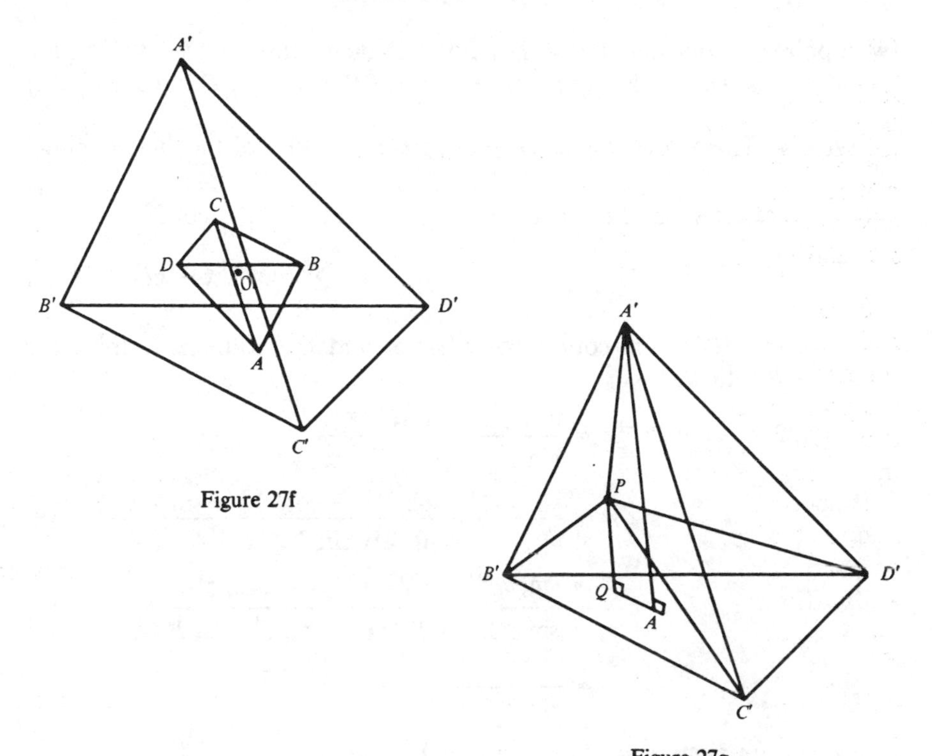

Figure 27f

Figure 27g

Third solution. Suppose $A'B'C'D'$ is a regular tetrahedron (See Fig. 27f). Let $A'A$, $B'B$, $C'C$ and $D'D$ be the altitudes from its vertices to the opposite faces. Then (by symmetry) A, B, C, D are the vertices of a regular tetrahedron inscribed in $A'B'C'D'$. Moreover, the four altitudes $A'A$, $B'B$, $C'C$, and $D'D$ intersect in a point O which is the circumcenter of both tetrahedra. If P is any point in $A'B'C'D'$, see Fig. 27g, clearly $PA + PB + PC + PD$ is $\geqslant$ the sum $s = PQ + PR + PS + PT$ of the altitudes from P to the four faces of $A'B'C'D'$.

We assert that s is independent of P. To see this, note that the volume V of $A'B'C'D'$ is equal to the sum of the volumes of the four tetrahedra into which it is divided by P (Fig. 27g). These four tetrahedra have bases of equal area $\triangle$, so $V = \frac{1}{3}(PQ + PR + PS + PT)\triangle = \frac{1}{3}s\triangle$. Thus $s = 3V/\triangle$, showing that s is independent of P.

In particular, taking $P = O$, we find that $s = OA + OB + OC + OD$. Hence for any point P in $A'B'C'D'$, we have

$$PA + PB + PC + PD \geqslant OA + OB + OC + OD\ .$$

Clearly equality occurs only if $PA \perp B'C'D'$ etc. This means that P lies on the lines AA', BB', CC' and DD', i.e. $P = O$. (There is no problem with points P outside $A'B'C'D'$, for the above reasoning shows that for such points $s\triangle > 3V$, and hence $PA + PB + PC + PD \geqslant s > 3V/\triangle$.)

1966/4. The expression $\cot x - \cot 2^n x$ is the sum of the "telescoping" series

$$\cot x - \cot 2x + \cot 2x - \cot 2^2x + \cdots + \cot 2^{n-1}x - \cot 2^n x$$

$$= \sum_{k=0}^{n-1} (\cot 2^k x - \cot 2^{k+1}x) .$$

We use the definition $\cot\theta = \cos\theta/\sin\theta$ and the addition formula for $\sin(\theta_1 - \theta_2)$ to write

$$\begin{aligned}\cot 2^k x - \cot 2^{k+1}x &= \frac{\cos 2^k x}{\sin 2^k x} - \frac{\cos 2^{k+1}x}{\sin 2^{k+1}x}\\ &= \frac{\sin 2^{k+1}x \cos 2^k x - \cos 2^{k+1}x \sin 2^k x}{\sin 2^k x \sin 2^{k+1}x}\\ &= \frac{\sin(2^{k+1}x - 2^k x)}{\sin 2^k x \sin 2^{k+1}x} = \frac{\sin 2^k x}{\sin 2^k x \sin 2^{k+1}x}\\ &= \frac{1}{\sin 2^{k+1}x} .\end{aligned}$$

We now sum from $k = 0$ to $k = n - 1$ to obtain

$$\sum_{k=0}^{n-1} (\cot 2^k x - \cot 2^{k+1}x) = \frac{1}{\sin 2x} + \frac{1}{\sin 4x} + \cdots + \frac{1}{\sin 2^n x} ,$$

the desired result.

Remark. This proof can be motivated as follows. Suppose we set out to prove that

$$(1) \qquad \frac{1}{\sin 2x} + \frac{1}{\sin 4x} + \cdots + \frac{1}{\sin 2^n x} = \cot x - \cot 2^n x$$

by mathematical induction on n. Clearly, (1) holds for $n = 1$, since then the right side of (1) is

$$\begin{aligned}\cot x - \cot 2x &= \frac{\cos x}{\sin x} - \frac{\cos 2x}{\sin 2x} = \frac{\cos x}{\sin x} - \frac{\cos^2 x - \sin^2 x}{2 \sin x \cos x}\\ &= \frac{1}{2 \sin x \cos x} = \frac{1}{\sin 2x} ,\end{aligned}$$

and this is equal to the left side of (1) when $n = 1$. We must prove that if (1) holds for some n, it also holds for $n + 1$. To do this, add $1/\sin 2^{n+1}x$ to both sides of (1). The right side becomes $\cot x - \cot 2^n x + 1/\sin 2^{n+1}x$. To complete the induction, we must show that this is equal to $\cot x - \cot 2^{n+1}x$, which is tantamount to proving that

$$\cot 2^n x - \cot 2^{n+1}x = \frac{1}{\sin 2^n x} .$$

This is carried out as above.

1966/5. The i-th equation is of the form

$$\sum_{j=1}^{4} |a_i - a_j|x_j = 1 , \qquad i = 1, 2, 3, 4,$$

and the system is unaltered under any permutation of the subscripts 1, 2, 3, 4. More precisely, if $x_1, \ldots, x_4$ is a solution of the system for some definite choice of $a_1, \ldots, a_4$, and if we permute the a_i's, then the same permutation, when applied to $x_1, \ldots, x_4$, yields a solution of the new system.

Since the a_i are distinct, we may therefore assume that

$$a_1 > a_2 > a_3 > a_4$$

and write the equations as

$$(1) \qquad 0 + (a_1 - a_2)x_2 + (a_1 - a_3)x_3 + (a_1 - a_4)x_4 = 1$$

$$(2) \qquad (a_1 - a_2)x_1 + 0 + (a_2 - a_3)x_3 + (a_2 - a_4)x_4 = 1$$

$$(3) \qquad (a_1 - a_3)x_1 + (a_2 - a_3)x_2 + 0 + (a_3 - a_4)x_4 = 1$$

$$(4) \qquad (a_1 - a_4)x_1 + (a_2 - a_4)x_2 + (a_3 - a_4)x_3 + 0 = 1 .$$

Probably the easiest way to solve them is to subtract each equation from the one following it. Thus the second minus the first yields

$$(a_1 - a_2)(x_1 - x_2 - x_3 - x_4) = 0 ,$$

and dividing by the positive factor $a_1 - a_2$, we find that

$$(5) \qquad x_1 - x_2 - x_3 - x_4 = 0 .$$

Similarly, (3) minus (2) yields

$$(6) \qquad x_1 + x_2 - x_3 - x_4 = 0 ,$$

and (4) minus (3) yields

$$(7) \qquad x_1 + x_2 + x_3 - x_4 = 0 .$$

Subtracting (5) from (6), we find that $x_2 = 0$, and subtracting (6) from (7), that $x_3 = 0$. Now any one of (5), (6) or (7) yields $x_1 = x_4$. Substituting these results into any of the four original equations, we obtain

$$x_1 = x_4 = \frac{1}{a_1 - a_4} , \qquad x_2 = x_3 = 0 .$$

Remark. The solution of n equations of the same form,

$$\sum_{j=1}^{n} |a_i - a_j|x_j = 1 , \qquad i = 1, 2, \ldots, n , \qquad a_1 > a_2 > \cdots > a_n ,$$

is achieved in exactly the same manner and yields

$$x_1 = x_n = \frac{1}{a_1 - a_n} , \qquad x_2 = x_3 = \cdots = x_{n-1} = 0 .$$

If the arrangement according to size of the a_i is different, say $a_{k_1} > a_{k_2} > \cdots > a_{k_n}$, where $k_1, k_2, \ldots, k_n$ is some permutation of $1, 2, \ldots, n$, then the solution is

$$x_{k_1} = x_{k_n} = \frac{1}{a_{k_1} - a_{k_n}}, \qquad x_{k_2} = x_{k_3} = \cdots = x_{k_{n-1}} = 0 .$$

1966/6 First solution. Let A', B', C' be the midpoints of sides BC, CA and AB, respectively, and denote the sides of the "inner" triangle, $\triangle A'B'C'$, by a', b', c' (see Figures 28). $\triangle ABC$ is divided into four congruent triangles, each having area $\frac{1}{4}(ABC)$. Now let K, L, M be arbitrary points on BC, CA and AB, respectively. We shall prove that the area of at least one of the "outer" triangles AML, BKM, CLK is $\leqslant \frac{1}{4}(ABC)$ by first considering

A: the case where at least one of the segments LM, MK, KL does not cross one of the sides a', b', c' of the inner triangle (Figures 28 a, b), and then

B: the case where ML crosses a', MK crosses b', and KL crosses c' (Figure 28c).

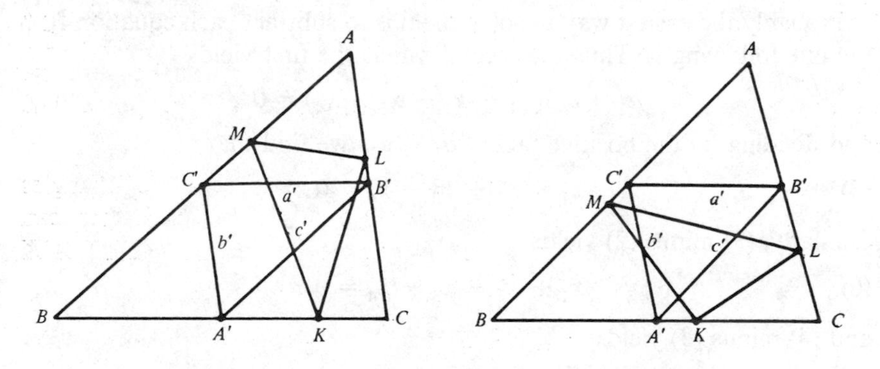

Figure 28a Figure 28b

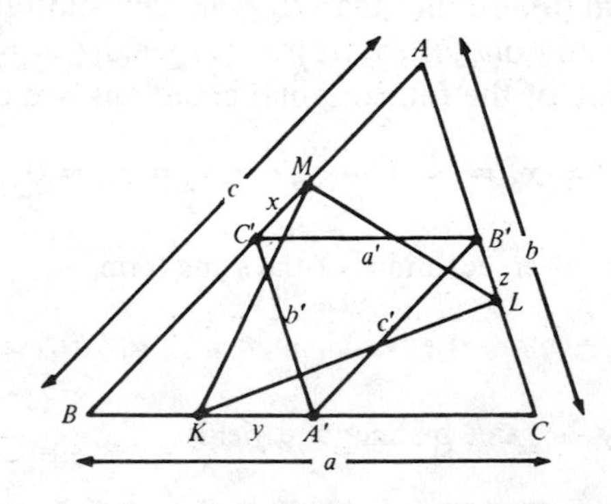

Figure 28c

A. Suppose ML does not cross a'. Then ML either lies outside $\triangle A'B'C'$ and $(AML) < (A'C'B') = \frac{1}{4}(ABC)$ (see Fig. 28a), or ML cuts both b' and c' (see Fig. 28 b). In the latter case either KL or MK lies outisde $\triangle A'B'C'$, depending on whether K is on segment $A'C$ or on segment BA'. So either

$$(CKL) < (CA'B') = \frac{1}{4}(ABC) \quad \text{or} \quad (BKM) < (BA'C') = \frac{1}{4}(ABC)\ .$$

Thus the assertion is proved in case A.

B. Figure 28c, in which ML crosses a', KL crosses c' and MK crosses b' is so drawn that ML extended meets BC extended, LK extended meets AB extended, and KM extended meets CA extended.

One way of proving the desired result is to show that the inner triangle KLM has area at least $\frac{1}{4}(ABC)$ (since the sum of the areas of all four is (ABC)). This is done by proving the relations

$$\frac{1}{4}(ABC) = (A'B'C') = (A'B'M) < (A'LM) < (KLM)\ .\dagger$$

The equality $(A'B'C') = (A'B'M)$ holds because $C'M \| A'B'$. The inequality $(A'B'M) < (A'LM)$ holds because the altitude from B' to $A'M$ is less than that from L to $A'M$ (since $A'M$ extended meets CA extended). Finally, the inequality $(A'LM) < (KLM)$ holds because the altitude from A' to ML is shorter than that from K to ML (since ML extended meets BC extended).

Another proof shows that the product of the areas of the outer triangles is $\leqslant [(ABC)/4]^3$, from which it follows that not all three factors can exceed $(ABC)/4$. This is accomplished by denoting segments $C'M$, $A'K$ and $B'L$ by x, y, z, respectively, and writing

$$2(AML) = \left(\frac{c}{2} - x\right)\left(\frac{b}{2} + z\right)\sin A\ ,$$

$$2(BKM) = \left(\frac{a}{2} - y\right)\left(\frac{c}{2} + x\right)\sin B\ ,$$

$$2(CKL) = \left(\frac{a}{2} + y\right)\left(\frac{b}{2} - z\right)\sin C\ .$$

The product of these equations is

$$\begin{aligned} &8(AML)(BKM)(CKL) \\ (1)\qquad &= \left(\frac{c^2}{4} - x^2\right)\left(\frac{a^2}{4} - y^2\right)\left(\frac{b^2}{4} - z^2\right)\sin A \sin B \sin C\ , \end{aligned}$$

while

$$2(ABC) = bc \sin A = ac \sin B = ab \sin C\ ;$$

†This proof is due to Russell Lyons.

so

$$(2) \qquad 8(ABC)^3 = a^2b^2c^2 \sin A \sin B \sin C\ .$$

The quotient of equations (1) and (2) is

$$(3) \qquad \frac{(AML)(BKM)(CKL)}{(ABC)^3} = \left(\frac{1}{4} - \frac{x^2}{c^2}\right)\left(\frac{1}{4} - \frac{y^2}{a^2}\right)\left(\frac{1}{4} - \frac{z^2}{b^2}\right).$$

Now $0 \leqslant x < c/2$, $0 \leqslant y < a/2$, $0 \leqslant z < b/2$, see Fig. 28c; therefore each of the numbers x^2/c^2, y^2/a^2, z^2/b^2 is less than $1/4$. It follows that the right side of (3) is $\leqslant (1/4)^3$, so

$$(4) \qquad (AML)(BKM)(CKL) \leqslant \left[\frac{(ABC)}{4}\right]^3.$$

This implies that at least one of the outer triangles AML, BKM, CKL has area $\leqslant (ABC)/4$, which is what we wished to prove. If K, L, M are the midpoints A', B', C', then $x = y = z = 0$, and (4) becomes the equality mentioned at the beginning of the problem.

Remark. We actually showed that in case B the area of the inner triangle is least when its vertices are the midpoints of the sides of $\triangle ABC$.

Second solution. Perform an affine transformation† sending $\triangle ABC$ onto an equilateral triangle, $\triangle A'B'C'$ (Fig. 28d). Such a transformation preserves ratios of areas; thus for example

$$\frac{(AML)}{(ABC)} = \frac{(A'M'L')}{(A'B'C')}\ .$$

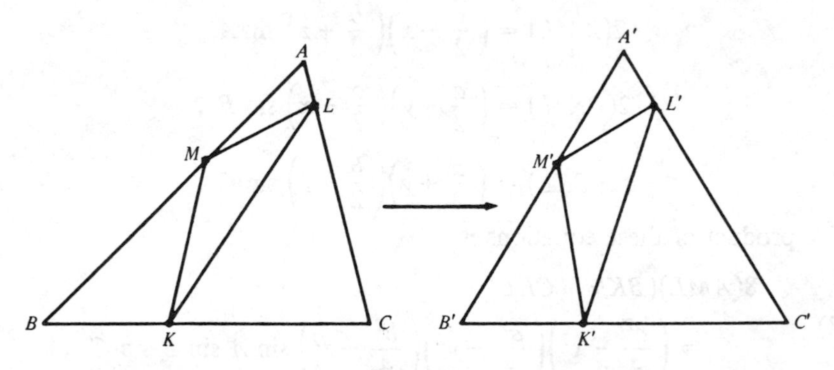

Figure 28d

†For a definition of affine transformations and a list of their properties, see e.g. *Geometric Transformations* III, by I. M. Yaglom, NML 24, 1973, p. 9ff.

Hence if the theorem holds for $\triangle A'B'C'$, then it also holds for $\triangle ABC$. We may therefore suppose from the outset that $\triangle ABC$ is equilateral with side s. (See Fig. 28e.)

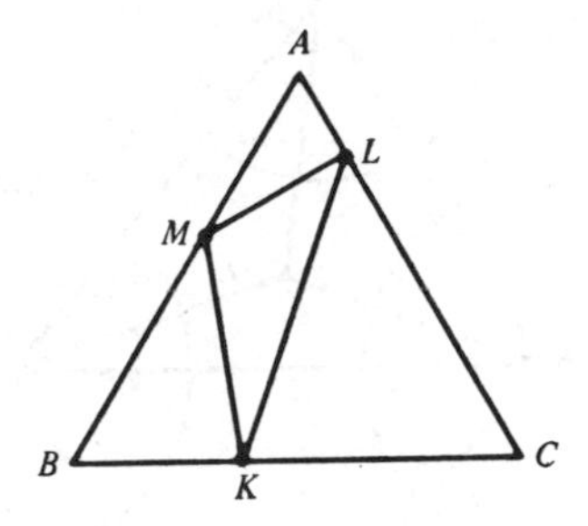

Figure 28e

Suppose without loss of generality that AL is the shortest of the six segments into which the sides of $\triangle ABC$ are divided by the points K, L, M. Then $AL \leqslant MB$, so $AM \cdot AL \leqslant AM \cdot MB$. By the arithmetic-geometric mean inequality,

$$AM \cdot MB \leqslant \left(\frac{AM + MB}{2} \right)^2 = \frac{1}{4} s^2 .$$

Hence

$$(AML) = \frac{1}{2} AM \cdot AL \sin 60° \leqslant \frac{1}{8} s^2 \sin 60° = \frac{1}{4} (ABC) .$$

Ninth International Olympiad, 1967

1967/1. Clearly, the unit circles centered at the vertices cover the parallelogram if and only if the unit circles centered at A, B, D cover $\triangle ABD$. To see when this happens, we first prove the following lemma:

LEMMA. *Let $\triangle ABD$ be an acute triangle, and let r be the radius of its circumscribed circle. Then the three circles of radius s centered at A, B, D cover $\triangle ABD$ if and only if $s \geqslant r$.*

PROOF: Since $\triangle ABD$ is acute, its circumcenter O lies inside the triangle. The distances OA, OB, and OD are all equal to r, so if $s < r$, O does not lie in any of the three circles of radius s centered at A, B, and D (see Fig. 29a). It therefore remains only to prove that the circles of radius r centered at A, B, D do indeed cover the triangle. (Circles of radius $> r$ then certainly will, since their union includes the union of the circles of radius r).

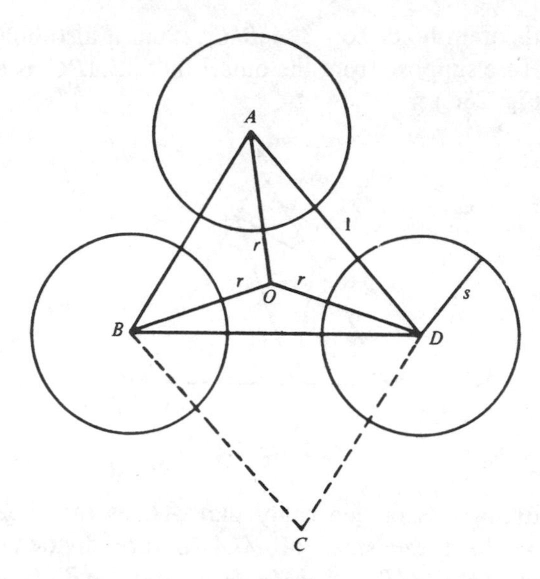

Figure 29a

Since the three circles of radius r intersect at O, it suffices to prove the more general result that if P is any point inside $\triangle ABD$, the three circles centered at A, B, D and passing through P cover the triangle. To show this, consider Fig. 29b. Let L, M, N be the feet of the perpendiculars from P to the sides BD, DA, AB respectively. Then $AN < AP$, since AN is the perpendicular from A to the line NP. Similarly, $AM < AP$. Hence the quadrilateral $AMPN$ lies inside the circle through P centered at A. Similarly, the quadrilaterals $BLPN$ and $DLPM$ lie inside the circles through P centered at B and at D respectively. It follows that $\triangle ABD$, the union of the three quadrilaterals, is contained in the union of the three circles. This completes the proof of the lemma.

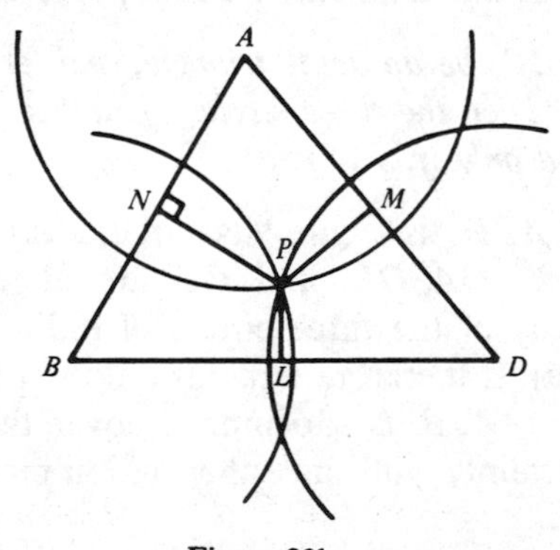

Figure 29b

It is an immediate consequence of the lemma that the unit circles centered at A, B and D cover $\triangle ABD$ if and only if $1 \geqslant r$. We shall now show that this condition is equivalent to $a \leqslant \cos\alpha + \sqrt{3}\sin\alpha$.

Let d denote the length of side BD (see Fig. 29c). By the law of cosines,

$$d^2 = 1 + a^2 - 2a\cos\alpha \ . \tag{1}$$

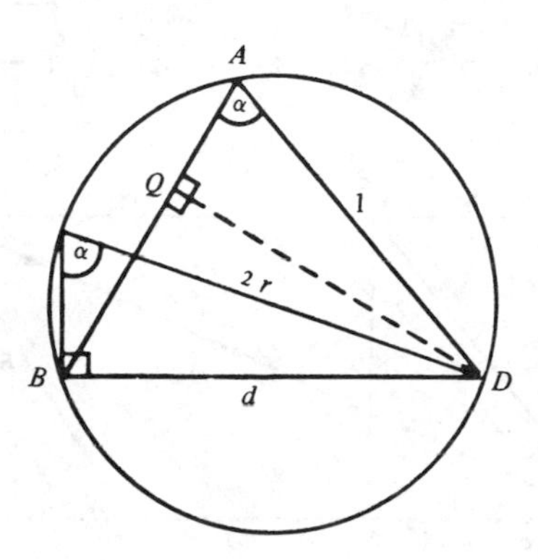

Figure 29c

On the other hand, we see from this figure that $d/2r = \sin\alpha$, and hence $d^2 = 4r^2\sin^2\alpha$. Substituting this into (1), we obtain

$$4r^2\sin^2\alpha = 1 + a^2 - 2a\cos\alpha \ . \tag{2}$$

Therefore $r \leqslant 1$ if and only if

$$4\sin^2\alpha \geqslant 1 + a^2 - 2a\cos\alpha \ . \tag{3}$$

On the right side of (3), replace the term 1 by $\cos^2\alpha + \sin^2\alpha$. Then (3) becomes equivalent to

$$3\sin^2\alpha \geqslant a^2 - 2a\cos\alpha + \cos^2\alpha = (a - \cos\alpha)^2 \ .$$

This in turn is equivalent to

$$\sqrt{3}\sin\alpha \geqslant |a - \cos\alpha| \ .$$

Since the condition in the statement of the problem is

$$\sqrt{3}\sin\alpha \geqslant a - \cos\alpha \ ,$$

it remains only to show that $a - \cos\alpha \geqslant 0$. To do this, we draw the altitude DQ from D to AB (see Fig. 29d). Since $\triangle ABD$ is acute, Q lies inside the segment AB, so $AQ < AB$. But $AQ = \cos\alpha$ and $AB = a$, so $\cos\alpha < a$. This completes the solution.

1967/2. Let the vertices of the tetrahedron be A, B, C, D with all edges $\leqslant 1$ except AD. Denote the sides of $\triangle ABC$ by a, b, c, and the altitude from A to BC by h. Through the midpoint M of BC, draw a perpendicular MN of length h. (See Fig. 30.) Suppose for definiteness that A and C lie on the same side of the line MN. Then $BN \leqslant BA$; squaring both sides, we get $h^2 + (a/2)^2 \leqslant c^2$. Since by hypothesis $c \leqslant 1$, this implies that

$$h^2 \leqslant 1 - \frac{a^2}{4}, \quad \text{so} \quad h \leqslant \sqrt{1 - \frac{a^2}{4}}\ .$$

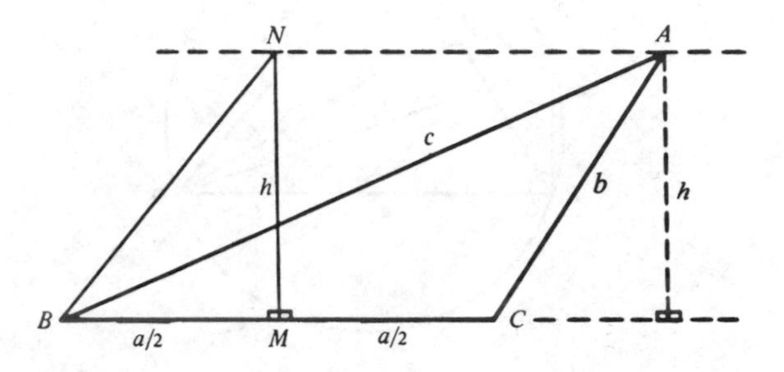

Figure 30

Let k be the altitude from D to BC in $\triangle DBC$, whose sides also have lengths $\leqslant 1$. The same analysis shows that $k \leqslant \sqrt{1 - (a^2/4)}$. The altitude of the tetrahedron from D to base $\triangle ABC$ is at most k, so the volume V of $ABCD$ is

$$\frac{1}{3}\left(\frac{1}{2}ah\right)k \leqslant \frac{1}{6}a\sqrt{1 - \frac{a^2}{4}}\sqrt{1 - \frac{a^2}{4}}$$
$$= \frac{1}{6}a\left(1 + \frac{a}{2}\right)\left(1 - \frac{a}{2}\right).$$

But $a \leqslant 1$, so

$$V \leqslant \frac{1}{6}a\frac{3}{2}\left(1 - \frac{a}{2}\right) = \frac{1}{8}\left[2a - a^2\right] = \frac{1}{8}\left[1 - (a-1)^2\right] \leqslant \frac{1}{8}\ .$$

Equality holds if $\triangle ABC$ and $\triangle DBC$ are equilateral with sides of length 1, lying in perpendicular planes.

1967/3. By the definition of c_s, we have

$$c_a - c_b = a(a+1) - b(b+1) = a^2 - b^2 + a - b$$
$$= (a-b)(a+b+1)\ .$$

Hence the factors of the first product are

$$c_{m+1} - c_k = (m+1-k)(m+k+2)$$
$$c_{m+2} - c_k = (m+2-k)(m+k+3)$$
$$\cdots\cdots\cdots\cdots\cdots\cdots\cdots\cdots$$
$$c_{m+n} - c_k = (m+n-k)(m+k+n+1) \; .$$

Their product is

$$[(m-k+1)(m-k+2)\cdots(m-k+n)]$$
$$\times[(m+k+2)(m+k+3)\cdots(m+k+n+1)] \; ,$$

which we write as $A \cdot B$, where A is a product of n consecutive integers starting with $m-k+1$, and B is a product of n consecutive integers starting with $m+k+2$. Now the factors in the product $c_1c_2\cdots c_n$ are

$$c_1 = 1\cdot 2, \qquad c_2 = 2\cdot 3, \qquad \cdots, \qquad c_n = n(n+1) \; ,$$

and their product is

$$n!\,(n+1)! \; .$$

We shall solve the problem by showing that A is divisible by $n!$, and B is divisible by $(n+1)!$.

Let us assume for a moment that the product of *any* n consecutive integers is divisible by $n!$, a fact that we shall establish at the end of this solution. It follows at once that A is divisible by $n!$. It also follows that the product $(m+k+1)B$ of $n+1$ consecutive integers is divisible by $(n+1)!$. But we are told that $m+k+1$ is a prime greater than $n+1$, so it is relatively prime to $(n+1)!$. Hence B is divisible by $(n+1)!$ as was to be shown.

We now complete the argument by proving:

LEMMA. *The product of any* n *consecutive integers is divisible by* $n!$.

PROOF: We consider three cases: the n consecutive integers (i) are all positive, (ii) are all negative, (iii) include zero. The third case is easy because the product then is zero, and zero is divisible by $n!$. In case (i), let the n consecutive positive integers be $j+1, j+2, \ldots, j+n$; we want to prove that

$$\frac{(j+1)(j+2)\cdots(j+n)}{n!} = \frac{(j+n)!}{j!\,n!}$$

is an integer. But this is an integer because it is a binomial coefficient $\binom{n+j}{n}$, i.e. the number of combinations of $n+j$ things n at a time. Case (ii) can be reduced to case (i) by replacing the negative integers in the product by their absolute values without affecting the divisibility properties of the product.

1967/4. Through A_0, B_0, C_0, draw lines parallel to B_1C_1, C_1A_1, and A_1B_1, respectively; see Figure 31a. These form the sides BC, CA and AB of a $\triangle ABC$ similar to $\triangle A_1B_1C_1$.

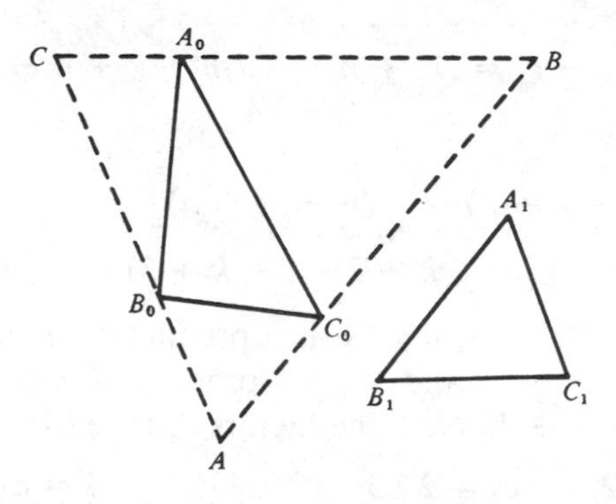

Figure 31a

Now suppose each of the lines just drawn is rotated about A_0, B_0, C_0, respectively, by the same amount; then they meet at the same angles as before, always forming triangles similar to $\triangle A_1B_1C_1$. The triangle of maximum area among them is the one whose sides have maximal length. To find it, recall that the locus of all points B such that $\angle A_0BC_0$ has a given measure β is an arc of a circle with chord A_0C_0. This suggests that we construct the circumcircles of triangles A_0C_0B, B_0A_0C and B_0C_0A. Denote their centers by O_b, O_c and O_a, respectively, see Fig. 31b.

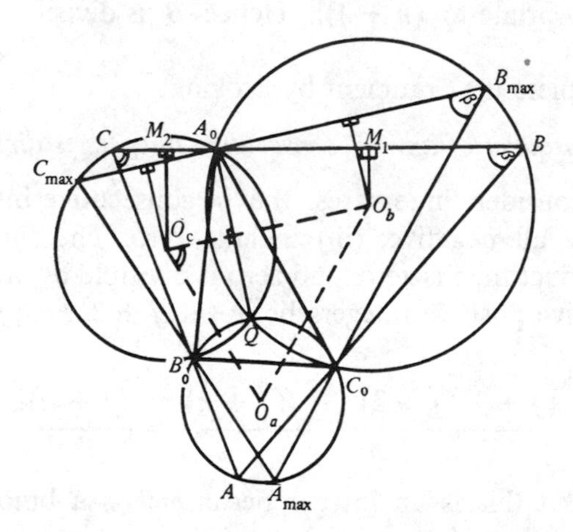

Figure 31b

It is easy to prove that these circumcircles have a point Q in common. [This is a consequence of the fact that $\angle A + \angle B + \angle C = 180°$; for a proof, see e.g. *Geometry Revisited*, by H. S. M. Coxeter and S. L. Greitzer, vol. 19 in this NML series, Theorem 3.31 on p. 61.]

We show next that $\triangle O_aO_bO_c \sim \triangle ABC$.

PROOF: $\angle C = \frac{1}{2}\widehat{A_0QB_0}$; and $\angle O_aO_cO_b = \frac{1}{2}\widehat{A_0Q} + \frac{1}{2}\widehat{QB_0}$, because O_cO_a and O_cO_b bisect arcs B_0Q and A_0Q, respectively. So $\angle C = \angle O_aO_cO_b$. Similarly, $\angle A = \angle O_cO_aO_b$, $\angle B = \angle O_aO_bO_c$. Therefore $\triangle O_aO_bO_c \sim \triangle ABC \sim \triangle A_1B_1C_1$.

Finally, we show that the largest triangle ABC through points A_0, B_0, C_0 is the one whose sides are parallel to those of $\triangle O_aO_bO_c$.

PROOF: Since perpendiculars from O_b and O_c bisect the chords BA_0 and CA_0 at M_1 and M_2, we have $M_1M_2 = \frac{1}{2}BC$. M_1M_2 is the perpendicular projection of O_bO_c on BC and is largest when $BC \| O_bO_c$. Since $\triangle O_aO_bO_c \sim \triangle ABC$, all three sides of the maximal triangle are parallel to those of $\triangle O_aO_bO_c$.

Thus, to construct the maximal triangle, first construct any triangle through A_0, B_0, C_0 similar to $\triangle A_1B_1C_1$ (see first paragraph of this solution). Then construct the centers O_a, O_b, O_c of the circumcircles of triangles AB_0C_0, BA_0C_0, CB_0A_0, and finally construct lines through A_0, B_0, C_0 parallel to O_bO_c, O_cO_a and O_aO_b, respectively. They form the sides BC, CA and AB of the desired maximal triangle.

1967/5. Assume the quantities are numbered so that $|a_1| \geqslant |a_2| \geqslant \cdots \geqslant |a_8|$. We may even assume, without loss of generality, that $a_1 = 1$, because $a_1 \neq 0$, and

$$\frac{c_n}{a_1^n} = 1^n + \left(\frac{a_2}{a_1}\right)^n + \cdots + \left(\frac{a_8}{a_1}\right)^n$$

vanishes for the same set of indices as c_n. So consider

(1) $\quad c_n = 1 + a_2^n + a_3^n + \cdots + a_8^n\,, \qquad \text{where } 1 \geqslant |a_2| \geqslant \cdots \geqslant |a_8|\,.$

For even values of n, $c_n > 0$, since no term of (1) is negative in this case. Therefore, c_n must be 0 for infinitely many odd values of n.

Suppose k of the a's (including $a_1 = 1$) are equal to 1, and l of them are equal to -1. Then

(2) $$c_{2m+1} = k - l + \sum_{i=k+l+1}^{8} a_i^{2m+1}\,.$$

Since the a's in the sum are less than 1 in absolute value, the sum tends to zero as m tends to infinity, so $c_{2m+1} \to k - l$. Hence c_{2m+1} can vanish infinitely often only if $k = l$. This shows that the terms of largest absolute value occur in pairs of opposite sign. From (2) we now get $\Sigma_{i=k+l+1}^{8} a_i^{2m+1} = 0$ infinitely often. Proceeding as above, we find that the terms with second largest absolute value occur in pairs of opposite sign, etc.

We conclude: if $c_n = 0$ holds for infinitely many n, then it holds for all odd values of n.

1967/6 First solution. Let u_k be the number of medals left at the beginning of the k-th day of the contest. Then the number distributed that day is

$$k + \tfrac{1}{7}(u_k - k)\,, \qquad \text{and} \qquad u_k - \left[k + \tfrac{1}{7}(u_k - k)\right] = \tfrac{6}{7}(u_k - k)$$

remain. So, at the beginning of day $k + 1$, there are

$$u_{k+1} = \tfrac{6}{7}(u_k - k)$$

medals left. Solving this for u_k we obtain

$$(1) \qquad u_k = \tfrac{7}{6}u_{k+1} + k \qquad \text{or} \qquad u_k - \tfrac{7}{6}u_{k+1} = k\,.$$

We are told also that $u_n = n$ and $u_1 = m$. This together with equations (1) leads to the system

$$\begin{aligned} m - (7/6)u_2 &= 1 \\ u_2 - (7/6)u_3 &= 2 \\ u_3 - (7/6)u_4 &= 3 \\ \cdots\cdots\cdots &= \cdots \\ u_{n-2} - (7/6)u_{n-1} &= n-2 \\ u_{n-1} - (7/6)n &= n-1\,. \end{aligned}$$

Multiply these equations by the factors $1, (7/6), (7/6)^2, \ldots, (7/6)^{n-2}$, and then add. The sum of the left sides telescopes to $m - (7/6)^{n-1}n$, so we obtain

$$(2) \qquad m = 1 + 2(\tfrac{7}{6}) + 3(\tfrac{7}{6})^2 + \cdots + (n-1)(\tfrac{7}{6})^{n-2} + n(\tfrac{7}{6})^{n-1}\,.$$

The problem now is to find all natural numbers $n \geqslant 2$ for which the right side is an integer. Multiplying (2) by $7/6$, we obtain

$$\tfrac{7}{6}m = \tfrac{7}{6} + 2(\tfrac{7}{6})^2 + 3(\tfrac{7}{6})^3 + \cdots + n(\tfrac{7}{6})^n\,,$$

and subtracting this from (2) yields

$$(3) \qquad -\tfrac{1}{6}m = [1 + \tfrac{7}{6} + (\tfrac{7}{6})^2 + \cdots + (\tfrac{7}{6})^{n-1}] - n(\tfrac{7}{6})^n\,.$$

The sum in the square brackets is a geometric progression with value

$$\frac{(7/6)^n - 1}{(7/6) - 1} = 6\left[\left(\frac{7}{6}\right)^n - 1\right].$$

Substituting this in (3) and multiplying by -6, we get

$$m = -36\left[\left(\frac{7}{6}\right)^n - 1\right] + 6n\left(\frac{7}{6}\right)^n = 36 - \frac{(n-6)7^n}{6^{n-1}} .$$

This is an integer if and only if 6^{n-1} divides $(n-6)7^n$; so 6^{n-1} must divide $n-6$. But for $n > 1$ we have $6^{n-1} > |n-6|$, and so $n - 6 = 0$ is the only possibility. This gives $n = 6$ and $m = 36$. Thus the contest lasted 6 days, and 36 medals were awarded.

Second solution. Another way of solving the problem is to write (1) in the form

$$u_{k+1} - \tfrac{6}{7}u_k = -\tfrac{6}{7}k, \tag{4}$$

and use the theory of difference equations.† We look for a solution of the form $u_k = v_k + w_k$, where v_k is a particular solution of (4), and w_k is a solution of the corresponding homogeneous equation $w_{k+1} - \frac{6}{7}w_k = 0$.

To get a particular solution v_k, we try setting $v_k = Ak + B$. Then $v_{k+1} = A(k+1) + B$, and

$$v_{k+1} - \tfrac{6}{7}v_k = A(k+1) + B - \tfrac{6}{7}(Ak+B) = \tfrac{1}{7}Ak + A + \tfrac{1}{7}B = -\tfrac{6}{7}k .$$

Equating the coefficients of k, and then the constant terms, we find that

$$\frac{A}{7} = -\frac{6}{7}, \quad A = -6 ; \qquad \text{and} \quad A + \frac{B}{7} = -6 + \frac{B}{7} = 0 , \quad B = 42 ,$$

so

$$v_k = 42 - 6k .$$

For the homogeneous equation, we have

$$w_{k+1} = \tfrac{6}{7}w_k = \left(\tfrac{6}{7}\right)\left(\tfrac{6}{7}w_{k-1}\right) = \left(\tfrac{6}{7}\right)^2 w_{k-1} = \left(\tfrac{6}{7}\right)^3 w_{k-2} = \cdots = \left(\tfrac{6}{7}\right)^k w_1 .$$

Thus the complete solution of (4) is of the form

$$u_k = v_k + w_k = 42 - 6k + \left(\tfrac{6}{7}\right)^{k-1} w_1 .$$

†See for example S. Goldberg, *Introduction to Difference Equations*, John Wiley and Sons, 1958.

On the $(n + 1)$st day, there are no medals left, so

$$(5) \qquad \begin{aligned} 0 = u_{n+1} = v_{n+1} + w_{n+1} &= 42 - 6(n + 1) + (\tfrac{6}{7})^n w_1 \\ &= 36 - 6n + (\tfrac{6}{7})^n w_1 \ . \end{aligned}$$

On the first day there are m medals, so

$$m = u_1 = v_1 + w_1 = 42 - 6 + w_1 = 36 + w_1 \ , \quad w_1 = m - 36 \ .$$

Substituting for w_1 in (5) yields $0 = 36 - 6n + (6/7)^n(m - 36)$, so

$$m = 6(n - 6)\left(\frac{7}{6}\right)^n + 36 = \frac{7^n(n - 6)}{6^{n-1}} + 36 \ .$$

This is precisely the equation we obtained in the first solution, and which, by virtue of the fact that m and n are natural numbers, yielded the unique answer $m = 36, n = 6$.

Tenth International Olympiad, 1968

1968/1. Denote the lengths of the sides by $b - 1$, b, $b + 1$ and the opposite angles by α, β, γ; see Figure 32. Obviously $b > 2$ and $\alpha < \beta < \gamma$. Using the law of cosines, we find that

$$(1) \qquad \cos \alpha = \frac{b^2 + (b + 1)^2 - (b - 1)^2}{2b(b + 1)} = \frac{b + 4}{2(b + 1)} \ ;$$

similarly

$$(2) \qquad \cos \beta = \frac{b^2 + 2}{2(b^2 - 1)} \ , \qquad \cos \gamma = \frac{b - 4}{2(b - 1)} \ .$$

Note that these fractions are rational numbers for any integer b. As b increases, $\cos \alpha$ decreases (the denominator in (1) growing twice as fast as the numerator), and so α increases. For $b \geqslant 7$ we see that $\cos \alpha \leqslant 11/16 < \sqrt{2}/2$, so $\alpha > 45°$. But then $\beta > 45°$, so $\gamma < 90°$, and no angle is twice another. We therefore need look only at $b = 3, 4, 5, 6$.

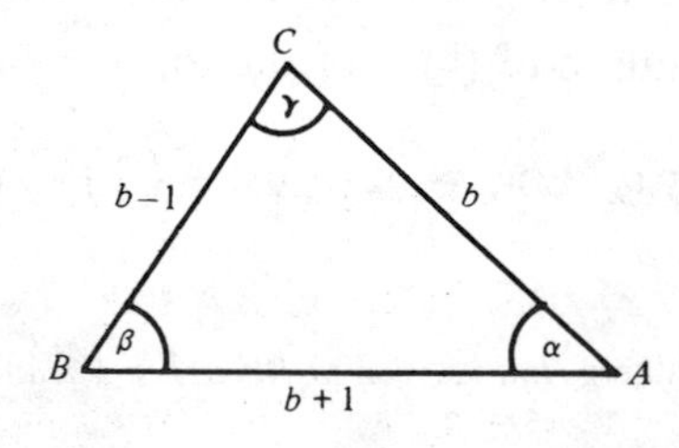

Figure 32

Now if $\gamma = 2\alpha$, $\gamma = 2\beta$, or $\beta = 2\alpha$, we have respectively

$$\cos\gamma = 2\cos^2\alpha - 1 \quad \text{or} \quad \cos\gamma = 2\cos^2\beta - 1 \quad \text{or}$$
$$\cos\beta = 2\cos^2\alpha - 1\ ,$$

and so

$$\cos\alpha = \sqrt{(1+\cos\gamma)/2} \quad \text{or} \quad \cos\beta = \sqrt{(1+\cos\gamma)/2} \quad \text{or}$$
$$\cos\alpha = \sqrt{(1+\cos\beta)/2}\ .$$

For $\cos\alpha$ or $\cos\beta$ to be rational, $(1+\cos\gamma)/2$ or $(1+\cos\beta)/2$ must be the square of a rational number. But for $b = 3, 4, 5, 6$ the values, computed by (2), of $(1+\cos\beta)/2$ are $27/32, 4/5, 25/32, 27/35$; and those of $(1+\cos\gamma)/2$ are $3/8, 1/2, 9/16, 3/5$. The only one that meets the specifications is $9/16$, with $b = 5$ and $\cos\alpha = 3/4$. Hence the only triangle of the required sort has sides 4, 5, 6 with $\gamma = 2\alpha$.

Remark. Another solution examines the three possibilities $\beta = 2\alpha$, $\gamma = 2\alpha$, $\gamma = 2\beta$, then uses the law of sines (or cosines) and trigonometric identities to derive equations which b must satisfy and finds in the first case that $b = 2$ (contradicting $b > 2$), in the second that $b = 5$ (leading to the 4, 5, 6 solution triangle) and in the third case that b is irrational.

1968/2. Assume that the number x has n digits:

$$x = a_0 + a_1 10 + a_2 10^2 + \cdots + a_{n-1}10^{n-1}\ , \qquad a_i \leqslant 9\ .$$

Let $P(x)$ be the product of the digits; then

$$P(x) = a_0 a_1 a_2 \cdots a_{n-1} = x^2 - 10x - 22\ .$$

But

$$P(x) = (a_0 a_1 a_2 \cdots a_{n-2})a_{n-1} \leqslant 9^{n-1}a_{n-1} < 10^{n-1}a_{n-1} \leqslant x\ ,$$

so

$$x^2 - 10x - 22 < x, \text{ that is}, x^2 - 11x < 22\ .$$

Now

$$x^2 - 11x + \frac{121}{4} = \left(x - \frac{11}{2}\right)^2 < 22 + \frac{121}{4} = \frac{209}{4}\ ,$$

so

$$x - \frac{11}{2} < \frac{\sqrt{209}}{2} < \frac{15}{2}\ .$$

It follows that $x < 13$. That is, x either has one digit, or $x = 10, 11$, or 12.

If x has one digit, then $x = a_0$, and $P(x) = a_0 = x = x^2 - 10x - 22$, so $x^2 - 11x - 22 = 0$. But this equation has no integral solutions.

If x has two digits, we can easily test all three possibilities:

(i) If $x = 10$, $P(x) = 0$, $x^2 - 10x - 22 = -22 \neq 0$, so the given condition is not met.

(ii) If $x = 11$, $P(x) = 1$, $x^2 - 10x - 22 = -11 \neq 1$, and again the given condition is violated.

(iii) If $x = 12$, $P(x) = 2$, $x^2 - 10x - 22 = 2$, so $x = 12$ is the only solution.

Instead of testing the three candidates, we could have found all two-digit solutions by setting $x = a_0 + 10$. Then

$$x^2 - 10x - 22 = a_0^2 + 20a_0 + 100 - 10a_0 - 100 - 22 = P(x) = a_0 .$$

Hence

$$a_0^2 + 9a_0 - 22 = (a_0 + 11)(a_0 - 2) = 0 ,$$

which implies $a_0 = 2$, i.e. $x = 12$.

1968/3. We add all n equations and obtain

$$a\sum_1^n x_i^2 + b\sum_1^n x_i + nc = \sum_1^n x_i ,$$

or

$$a\sum_1^n x_i^2 + (b-1)\sum_1^n x_i + nc = 0 . \tag{1}$$

Now consider the quadratic function

$$Q(y) = ay^2 + (b-1)y + c . \tag{2}$$

If the sum of its values at n (not necessarily distinct) points $x_1, x_2, \ldots, x_n$ is zero, then

$$Q(x_1) + Q(x_2) + \cdots + Q(x_n) = 0 , \tag{3}$$

which is precisely equation (1). If Q vanishes at some point z, then $x_1 = x_2 = \cdots = x_n = z$ is a solution of the given system. Now Q vanishes if and only if its graph (which is a parabola) has a point in common with the x-axis. We write

$$\begin{aligned} Q(y) &= a\left[y^2 + \frac{b-1}{a}y + \frac{c}{a}\right] \\ &= a\left\{y^2 + \frac{b-1}{a}y + \frac{(b-1)^2}{4a^2} - \frac{(b-1)^2}{4a^2} + \frac{c}{a}\right\} \\ &= a\left\{\left(y + \frac{b-1}{2a}\right)^2 - \frac{(b-1)^2 - 4ac}{4a^2}\right\} . \end{aligned}$$

The first term in the braces is $\geqslant 0$, with equality only when $y = -(b - 1)/2a$. Hence if $a > 0$, Q has a minimum at $-(b-1)/2a$, and if $a < 0$, Q has a maximum there. If the minimum of Q lies above the x-axis (or the maximum below) then Q never vanishes. This is the case if

$$Q\left(-\frac{b-1}{2a}\right) = a\left(-\frac{(b-1)^2 - 4ac}{4a^2}\right)\begin{cases} > 0 & \text{for } a > 0 \\ < 0 & \text{for } a < 0 \end{cases},$$

i.e. if $\triangle = (b-1)^2 - 4ac < 0$. This shows that for $\triangle < 0$, $Q(y)$ is either positive for all y or negative for all y, so the sum on the left of (3) cannot be zero. Hence neither (1) nor the given system has a solution.

If $\triangle = 0$, then $Q(-\dfrac{b-1}{2a}) = 0$, the parabola is tangent to the x-axis, and Q vanishes only at $-(b-1)/2a$. Hence the only solution of (3) is $x_1 = x_2 = \cdots = x_n = -(b-1)/2a$. This also satisfies the original system and is therefore its only solution.

If $\triangle > 0$, the graph of Q crosses the x-axis, Q has two distinct zeros, say z_1 and z_2, and $x_i = z_1$ $(i = 1, 2, \ldots, n)$ is a solution of the given system, as is $x_i = z_2$.

1968/4. We recall that a triangle can be constructed with three given segments for sides if and only if the sum of the lengths of any two exceeds the length of the third. Equivalently, the three segments are *not* the sides of a triangle if and only if the longest of them is $\geqslant$ the sum of the other two.

Denote the vertices of any tetrahedron by A, B, C, D, and let AB be the longest side. Suppose there is no vertex such that the edges meeting there are the sides of a triangle. Consider vertex A with attached edges AB, AC, AD. Then $AB \geqslant AC + AD$ by the above remarks. Similarly, by considering vertex B, we conclude that $BA \geqslant BC + BD$. Adding these inequalities, we get

$$2AB \geqslant AC + BC + AD + BD .$$

But from the triangular faces ABC and ABD we get $AB < AC + BC$ and $AB < AD + BD$; and if we add these two inequalities we get

$$2AB < AC + BC + AD + BD ,$$

a contradiction.

1968/5. (a) The given equation shows that $f(x+a) \geqslant \frac{1}{2}$, and so $f(x) \geqslant \frac{1}{2}$ for all x. Hence if we put $g(x) = f(x) - \frac{1}{2}$, we have $g(x) \geqslant 0$ for all x. The given functional equation now becomes

$$g(x+a) = \sqrt{g(x) + \tfrac{1}{2} - [g(x)]^2 - g(x) - \tfrac{1}{4}} = \sqrt{\tfrac{1}{4} - [g(x)]^2} .$$

Squaring we get

(1) $$[g(x+a)]^2 = \tfrac{1}{4} - [g(x)]^2 \qquad \text{for all } x\ ,$$

and hence also

$$[g(x+2a)]^2 = \tfrac{1}{4} - [g(x+a)]^2\ ;$$

these two imply $[g(x+2a)]^2 = [g(x)]^2$. Since $g(x) \geqslant 0$ for all x, we can take square roots to get $g(x+2a) = g(x)$, whence

$$f(x+2a) - \tfrac{1}{2} = f(x) - \tfrac{1}{2},$$

and

$$f(x+2a) = f(x) \qquad \text{for all } x\ .$$

This shows that $f(x)$ is periodic with period $2a$.

(b) To find all solutions, we set $h(x) = 4[g(x)]^2 - \frac{1}{2}$ and write (1) in the form

(2) $$h(x+a) = -h(x).$$

Conversely, if $h(x) \geqslant \frac{1}{2}$ and satisfies (2), then $g(x)$ satisfies (1).

An example for $a = 1$ is furnished by the function

$$h(x) = \sin^2 \frac{\pi}{2} x - \frac{1}{2}$$

which satisfies (2) with $a = 1$. For this h, $g(x) = \frac{1}{2}|\sin(\pi x/2)|$ and

$$f(x) = \frac{1}{2}\left|\sin \frac{\pi}{2} x\right| + \frac{1}{2}\ .$$

Actually $h(x)$ can be defined arbitrarily in $0 \leqslant x < a$ subject only to the condition $|h(x)| \leqslant \frac{1}{2}$; then (2) defines h for all x.

1968/6 First solution. Consider the set of integers $\{1, 2, 3, \dots, n\}$. Of these, $[(n+1)/2]$ are odd. Next we count those divisible by 2, but not by 2^2 (i.e., every fourth number: $2, 6, 10, \dots$), and find that there are $[(n+2)/2^2]$ of them. In general, for any positive integral power 2^k of 2, there are $[(n+2^k)/2^{k+1}]$ integers in $\{1, 2, \dots, n\}$ divisible by 2^k but not by 2^{k+1}. Since every natural number is of the form $2^k m$ (m odd) for exactly one integer $k \geqslant 0$, the set $\{1, 2, \dots, n\}$ is the disjoint union of the sets whose elements we have just counted. Hence the total number of elements in all these sets is n. Thus

$$\sum_0^{\infty} \left[\frac{n+2^k}{2^{k+1}} \right] = n\ .$$

(If $2^k > n$, then $[(n+2^k)/2^{k+1}] = 0$, so the series $\Sigma_{k=0}^{\infty}[(n+2^k)/2^{k+1}]$ terminates when $k > \log n/\log 2$.)

Second solution. We shall show that for any real number n, the given sum is, in fact, finite, and that its value is $[n]$. The first assertion is true because if k is so large that $n < 2^k$, we have $n + 2^k < 2^k + 2^k = 2^{k+1}$, so $(n + 2^k)/2^{k+1} < 1$ and $[(n + 2^k)/2^{k+1}] = 0$. We prove the second assertion by expressing the given sum as a telescoping sum. To this end we first prove:

LEMMA. $\left[\frac{x}{2}\right] + \left[\frac{x+1}{2}\right] = [x]$.

PROOF: If x is increased by 2, both sides of the asserted identity clearly increase by 2. Hence it suffices to verify the identity in the interval $0 \leqslant x < 2$. In this interval $[x/2] = 0$, so we need only show that

$$\left[\frac{x+1}{2}\right] = [x] \quad \text{for} \quad 0 \leqslant x < 2 .$$

If $0 \leqslant x < 1$, we have $[(x + 1)/2] = [x] = 0$, while if $1 \leqslant x < 2$, we have $[(x + 1)/2] = [x] = 1$. This completes the proof of the lemma, which we shall use in the form

$$[x] - \left[\frac{x}{2}\right] = \left[\frac{x+1}{2}\right] .$$

In this identity, set x successively equal to $n, n/2, n/2^2, \ldots$. We obtain

$$[n] - \left[\frac{n}{2}\right] = \left[\frac{n+1}{2}\right]$$

$$\left[\frac{n}{2}\right] - \left[\frac{n}{2^2}\right] = \left[\frac{(n/2)+1}{2}\right] = \left[\frac{n+2}{2^2}\right]$$

$$\left[\frac{n}{2^2}\right] - \left[\frac{n}{2^3}\right] = \left[\frac{(n/2^2)+1}{2}\right] = \left[\frac{n+2^2}{2^3}\right]$$

. . . .

$$\left[\frac{n}{2^k}\right] - \left[\frac{n}{2^{k+1}}\right] = \left[\frac{n+2^k}{2^{k+1}}\right] .$$

The integer k in the last line is chosen so that $2^k \leqslant n < 2^{k+1}$; thus

$$\left[\frac{n}{2^k}\right] \neq 0 , \qquad \text{but} \qquad \left[\frac{n}{2^{k+1}}\right] = 0 .$$

Adding these equalities, we find that the left sides form a sum telescoping to $[n]$, since terms cancel in pairs, and the right sides form the sum given in the problem. Of course if n is a natural number, then $[n] = n$.

Note. This problem appears on page 83 of the third edition of *The Theory of Numbers* by Niven and Zuckerman. It can also be solved by first writing n in binary notation, and then evaluating the various "integer part" expressions.†

Eleventh International Olympiad, 1969

1969/1. We look for natural numbers a of a form that allows us to factor $n^4 + a$. We claim that

$$a = 4k^4 , \qquad k = 2, 3, \ldots$$

has this property; for adding and subtracting $4n^2k^2$ enables us to write $n^4 + a$ as a difference of two squares, and this can always be factored. Thus

$$\begin{aligned} z = n^4 + 4k^4 &= n^4 + 4n^2k^2 + 4k^4 - 4n^2k^2 \\ &= (n^2 + 2k^2)^2 - (2nk)^2 \\ &= (n^2 + 2k^2 + 2nk)(n^2 + 2k^2 - 2nk) . \end{aligned}$$

It remains to show that for all $n \geqslant 1$, both factors exceed 1 if $k \geqslant 2$. Observe that

$$n^2 + 2k^2 + 2nk \geqslant n^2 + 2k^2 - 2nk = (n - k)^2 + k^2 \geqslant k^2 \geqslant 4 .$$

We conclude that each factor in the expression for z is, in fact, at least 4 for infinitely many k, hence for infinitely many $a = 4k^4$, and so z has the required property.

1969/2. With the help of the addition formula

$$\cos(\alpha + \beta) = \cos\alpha\cos\beta - \sin\alpha\sin\beta ,$$

$f(x)$ may be written in the form

$$\begin{aligned} f(x) &= \sum_{k=1}^{n} \frac{1}{2^{k-1}} \cos(a_k + x) \\ &= \sum_{k=1}^{n} \frac{1}{2^{k-1}} (\cos a_k \cos x - \sin a_k \sin x) \\ &= \left(\sum_{k=1}^{n} \frac{1}{2^{k-1}} \cos a_k \right) \cos x - \left(\sum_{k=1}^{n} \frac{1}{2^{k-1}} \sin a_k \right) \sin x \\ &= A \cos x - B \sin x , \end{aligned}$$

where

$$A = \sum_{1}^{n} \frac{\cos a_k}{2^{k-1}} , \qquad B = \sum_{1}^{n} \frac{\sin a_k}{2^{k-1}}$$

†Such a solution was given by Paul Zeitz.

Now A and B cannot both be zero, for, if they were, then $f(x)$ would vanish identically. However,

$$f(-a_1) = 1 + \frac{1}{2}\cos(a_2 - a_1) + \frac{1}{4}\cos(a_3 - a_1) + \cdots + \frac{1}{2^{n-1}}\cos(a_n - a_1)$$

$$\geqslant 1 - \frac{1}{2} - \frac{1}{4} - \cdots - \frac{1}{2^{n-1}} = \frac{1}{2^{n-1}} > 0 .$$

We are given that $f(x_1) = A\cos x_1 - B\sin x_1 = 0$, and $f(x_2) = A\cos x_2 - B\sin x_2 = 0$. If $A \neq 0$, we have $\cot x_1 = \cot x_2 = B/A$, and hence $x_2 - x_1 = m\pi$. If $A = 0$, then $B \neq 0$ and $\sin x_1 = \sin x_2 = 0$. Again we find that $x_2 - x_1 = m\pi$ for some integer m.

1969/3. We denote the tetrahedron by T and consider each value of k separately.

$k = 1$ Let T have vertices A, B, C, P with $AB = BC = AC = BP = CP = 1$, $AP = a$. Consider the plane rhombus $ABQC$ (shown in Fig. 33a) with sides of length 1 and diagonal BC of length 1. Then $AQ = \sqrt{3}$. Now the point P, one unit away from B and from C, is on the intersection of the unit spheres centered at B and C; i.e. P is anywhere (except at A or Q) on the circle with diameter AQ, the plane of the circle being perpendicular to the plane of the rhombus. Therefore $0 < AP < \sqrt{3}$, so

$$0 < a < \sqrt{3} . \tag{1}$$

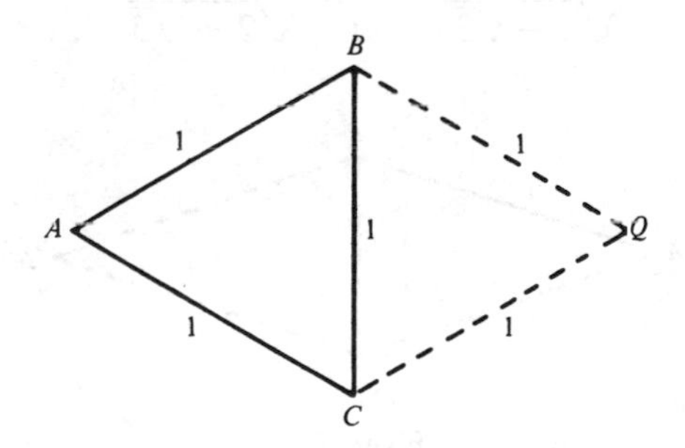

Figure 33a

$k = 2$ Here we distinguish two possible cases.

Case 1: The two edges of length a have a common vertex, say P, the other vertices of T being A, B, C. Suppose $PA = 1$, while $PB = PC = a$.

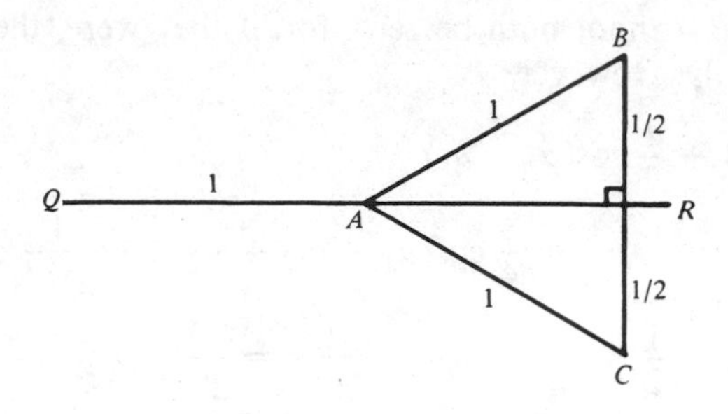

Figure 33b

Consider the equilateral triangle ABC with sides of length 1, and through A draw line QAR perpendicular to BC so that $AQ = AR = 1$; see Fig. 33b. Then

$$QB = QC = \sqrt{\left(\frac{1}{2}\right)^2 + \left(1 + \frac{\sqrt{3}}{2}\right)^2} = \sqrt{2+\sqrt{3}} \ ,$$

and

$$RB = RC = \sqrt{\left(\frac{1}{2}\right)^2 + \left(1 - \frac{\sqrt{3}}{2}\right)^2} = \sqrt{2-\sqrt{3}} \ .$$

Now P, being equidistant from B and C and one unit away from A lies on the intersection of the plane through QR perpendicular to $\triangle ABC$ with the unit sphere centered at A. Thus P can be anywhere on the circle with diameter QR and $\perp \triangle ABC$ (except at Q or R). Therefore PB and PC are greater than RB and less than QB. Thus in this case

$$\sqrt{2-\sqrt{3}} < a < \sqrt{2+\sqrt{3}} \ .$$

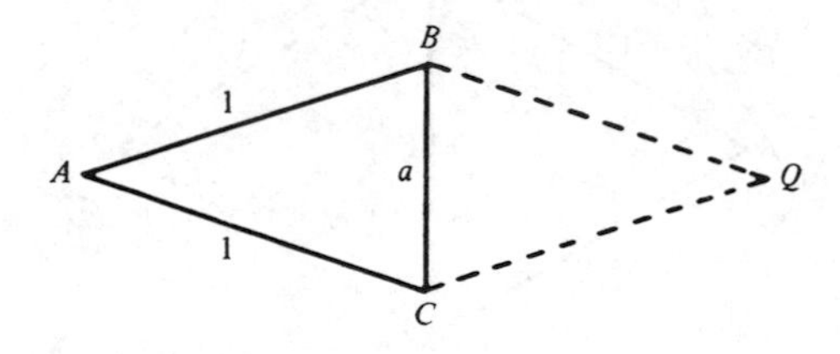

Figure 33c

Case 2: The two edges of length a are opposite; say $BC = AP = a$, and $AB = AC = PB = PC = 1$. Consider the plane rhombus $ABQC$ with sides of length 1 and diagonal $BC = a$; see Fig. 33c. Since $PB = PC = 1$, P lies on the intersection of the unit spheres centered at B and C, i.e. on the circle K with diameter AQ perpendicular to the rhombus. But $PA = a$, and hence a must be less than the diameter AQ of K. Since $AQ = 2\sqrt{1 - a^2/4}$, the condition $a < AQ$ amounts to $a^2 < 4 - a^2$, or $a < \sqrt{2}$.

From case 1 and case 2 we conclude that for $k = 2$, tetrahedra exist if and only if a lies in one of the intervals $(\sqrt{2-\sqrt{3}}, \sqrt{2+\sqrt{3}})$ or $(0, \sqrt{2})$, i.e.

$$(2) \qquad 0 < a < \sqrt{2+\sqrt{3}} .$$

$k = 3$ A tetrahedron with the specified edges always exists in this case. If $a \geqslant 1$, take $\triangle ABC$ with sides of length 1 and centroid Q. Let l be the line through Q perpendicular to $\triangle ABC$. Pick P on l so that $PA = PB = PC = a$. If $a \leqslant 1$, take $\triangle ABC$ with sides of length a, centroid Q, and line l as before. Pick P on l so that $PA = PB = PC = 1$.

$k = 4$ If four edges of T have length a and the remaining two have length 1, we consider a similar tetrahedron T' with four edges of length 1 and two of length $b = 1/a$. T' belongs to case $k = 2$; hence it exists if and only if $0 < b < \sqrt{2+\sqrt{3}}$, i.e.

$$a = \frac{1}{b} > \frac{1}{\sqrt{2+\sqrt{3}}} = \sqrt{2-\sqrt{3}} .$$

$k = 5$ Similarly, we reduce this to the case $k = 1$ and find, from (1), that the condition for T to exist is that

$$a > 1/\sqrt{3} .$$

We summarize our findings:

k	1	2	3	4	5
a	$0 < a < \sqrt{3}$	$0 < a < \sqrt{2+\sqrt{3}}$	$a > 0$	$a > \sqrt{2-\sqrt{3}}$	$a > \frac{1}{\sqrt{3}}$

1969/4. We shall show that the centers O_1 of γ_1, O_2 of γ_2 and O_3 of γ_3 are collinear. From this it follows immediately that the second common tangent is the mirror image of AB in the line through O_1, O_2 and O_3.

Suppose the incircle γ_1 of $\triangle ABC$ is tangent to AB at P, to BC at Q, and to CA at R; see Fig. 34a. Then $AR = AP$, $BQ = BP$, and $CR = CQ$. Denoting the lengths of the sides of $\triangle ABC$ by a, b, c, and its semi-perimeter $(a + b + c)/2$ by s, we find that

$$s - a = AP + BQ + CQ - (BQ + CQ) = AP .$$

Similarly $s - b = BP$ and $s - c = CQ = CR$. Thus

$$(1) \qquad AP = s - a , \qquad BP = s - b , \qquad CR = CQ = s - c .$$

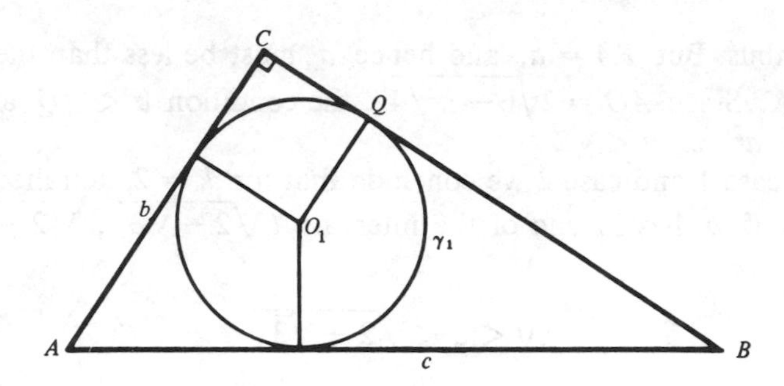

Figure 34a

Since $\angle C = 90°$, the points Q, C, R and O_1 from a square of side $s - c$, and the inradius is

(2) $$r_1 = s - c\ .$$

Fig. 34b shows circles γ_2 and γ_3, with centers O_2 and O_3 and radii r_2 and r_3. Let H_2 and H_3 be the feet of the perpendiculars from O_2 and O_3 to AB, and O is the center of γ. From the similar right triangles ABC, CBD and ACD, we get $(AC)^2 = (AB)(AD)$ and $(BC)^2 = (AB)(BD)$; hence

(3) $$AD = \frac{b^2}{c}\ , \qquad BD = \frac{a^2}{c}\ .$$

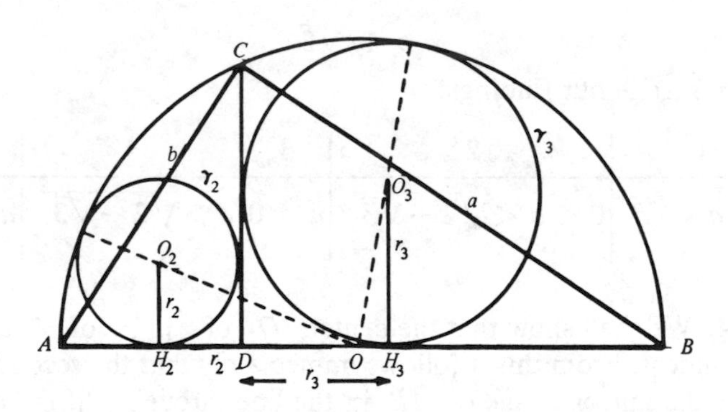

Figure 34b

To express r_2 in terms of a, b, c, we use the right triangle O_2H_2O, where $H_2O = r_2 + DO$ and $O_2O = r - r_2$. By the Pythagorean theorem,

$$r_2^2 + (r_2 + DO)^2 = (r - r_2)^2\ ,$$

which is equivalent to

$$r_2^2 + 2r_2(r + DO) = r^2 - (DO)^2 = (r + DO)(r - DO)$$

or $r_2^2 + 2r_2 BD = (BD)(AD)$, see Fig. 34b. In view of (3), this becomes

$$r_2^2 + 2r_2\frac{a^2}{c} = \frac{a^2}{c}\cdot\frac{b^2}{c} .$$

Adding a^4/c^2 to both sides yields

$$\left(r_2 + \frac{a^2}{c}\right)^2 = \frac{a^2(a^2 + b^2)}{c^2} = a^2 ,$$

whence

$$(4) \qquad r_2 + \frac{a^2}{c} = H_2B = a .$$

Similarly, we can find an analogous relation for r_3, the radius of circle γ_3, see Fig. 34b. From right $\triangle O_3H_3O$, where $H_3O = r_3 - DO$ and $O_3O = r - r_3$, we obtain

$$r_3^2 + (r_3 - DO)^2 = (r - r_3)^2$$

$$r_3^2 + 2r_3(r - DO) = (r + DO)(r - DO) ,$$

$$r_3^2 + 2r_3(AD) = (BD)(AD) .$$

Again we use (3), complete the square, and find that

$$(5) \qquad r_3 + \frac{b^2}{c} = AH_3 = b .$$

Adding (4) and (5), we find $r_2 + r_3 + (a^2 + b^2)/c = a + b$, or

$$(6) \qquad r_2 + r_3 = a + b - c = 2(s - c) .$$

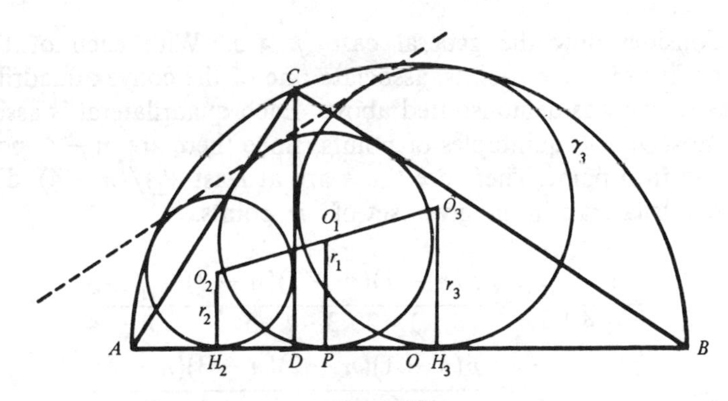

Figure 34c

In Fig. 34c, γ_1 is also drawn. We find that $H_2P = H_2B - PB$ which, by (1) and (4), yields $H_2P = a - (s - b) = a + b - s = s - c$, and

$H_3P = H_3A - AP = b - (s - a) = a + b - s = s - c$. Thus

$$H_2P = H_3P = s - c = r_1 = \tfrac{1}{2}(r_2 + r_3)\ . \tag{7}$$

It follows that O_1P is the midline of trapezoid $H_2H_3O_3O_2$, so O_1 lies at the midpoint of segment O_2O_3. This completes the proof of the collinearity of O_1, O_2, O_3.

1969/5 First solution. (i) Consider first the case $n = 5$: We must show that there is at least $\binom{5-3}{2} = 1$ convex quadrilateral. If the convex hull† of the five points has four of them on its boundary, they form a convex quadrilateral. If the boundary of the convex hull contains only three of the points, say A, B, C, then the other two, D and E, are inside $\triangle ABC$; see Figure 35a. Two of the points A, B, C must lie on the same side of the line DE. Suppose for definiteness that A and B lie on the same side of DE, as in Figure 35a. Then $ABDE$ is a convex quadrilateral.

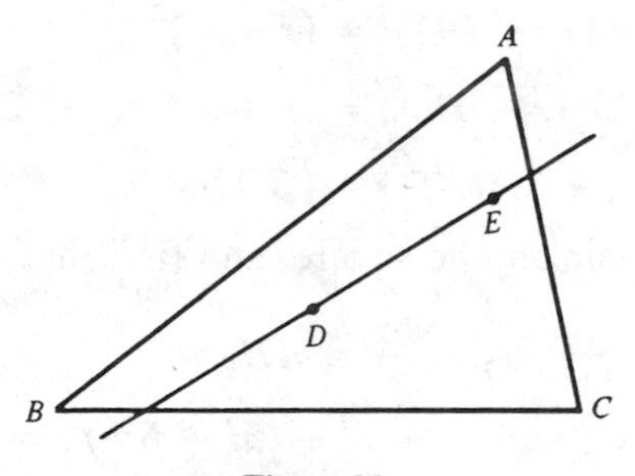

Figure 35a

(ii) Consider now the general case $n \geqslant 5$. With each of the $\binom{n}{5}$ subsets of five of the n points, associate one of the convex quadrilaterals whose existence was demonstrated above. Each quadrilateral is associated with at most $n - 4$ quintuples of points, since there are $n - 4$ possibilities for the fifth point. Therefore there are at least $\binom{n}{5}/(n - 4)$ different convex quadrilaterals in the given set of n points.

Now

$$\frac{1}{n-4}\binom{n}{5} = \frac{n(n-1)(n-2)(n-3)}{1\cdot 2\cdot 3\cdot 4\cdot 5}\cdot\frac{n-4}{n-4}$$

$$= \frac{n(n-1)(n-2)(n-3)(n-4)}{60(n-4)\qquad\qquad 1\cdot 2}$$

$$= \frac{n(n-1)(n-2)}{60(n-4)}\binom{n-3}{2}\ .$$

†The convex hull of a set S in the plane is the smallest convex set containing S.

Division of $n(n-1)(n-2)$ by $n-4$ yields $n^2+n+6+24/(n-4)$; when $n \geqslant 5$, $n^2+n+6+24/(n-4) \geqslant 60$, and hence

$$\frac{1}{n-4}\binom{n}{5} \geqslant \binom{n-3}{2} \qquad \text{for } n \geqslant 5 . \tag{1}$$

Inequality (1) may also be established from the fact that for $n \geqslant 5$, we have $n(n-1)(n-2) \geqslant 60(n-4)$. This in turn can be seen by forming the difference $n(n-1)(n-2)-60(n-4) = n^3-3n^2-58n+240 = (n-5)(n-6)(n+8)$, and observing that it vanishes for $n=5$ and $n=6$, and is positive for all greater n. So in (1) the equality sign holds only if $n=5$ or $n=6$.

Second solution. Choose three points A, B, C of the given set S which lie on the boundary of its convex hull. Then there are $\binom{n-3}{2}$ ways in which two additional points D and E can be selected from S. Once they are chosen, at least two of the points A, B, C must lie on the same side of the line DE. Suppose for definiteness that A and B are on the same side of DE. Then A, B, D, E are the vertices of a convex quadrilateral. For if not, their convex hull would be a triangle, as in Figure 35b. One of the points A, B would lie inside this triangle, contradicting the fact that A, B, C were chosen to be on the boundary of the convex hull of S. Thus we have found $\binom{n-3}{2}$ convex quadrilaterals whose vertices are among the given points.

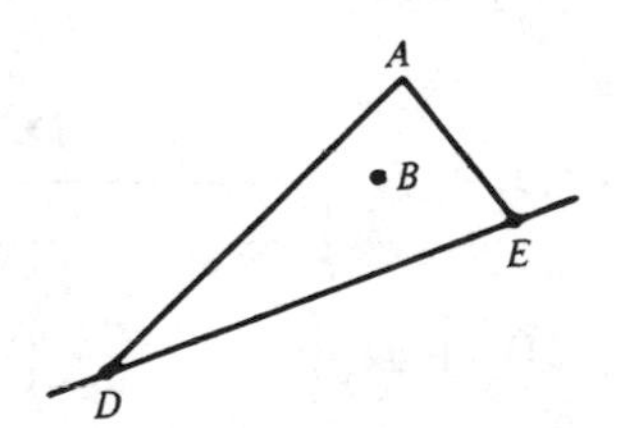

Figure 35b

Remark. The first solution is superior to the second in that it shows the existence of at least $\binom{n}{5}/(n-4)$ convex quadrilaterals, and this exceeds $\binom{n-3}{2}$ for all $n \geqslant 7$. (In fact $\binom{n}{5}/(n-4) \sim n^4/120$ as $n \to \infty$, while $\binom{n-3}{2} \sim n^2/2$.) However, in view of the way the problem was stated, it seems likely that the second solution is more along the lines the proposers had in mind.

1969/6. We introduce the abbreviations

$$\begin{aligned} D_1 &= x_1y_1 - z_1^2, \qquad D_2 = x_2y_2 - z_2^2, \\ D &= (x_1+x_2)(y_1+y_2) - (z_1+z_2)^2 \end{aligned} \tag{1}$$

and note that the given relations $D_i > 0, x_i > 0$ imply $y_i > 0$ $(i = 1, 2)$.

Also, by (1), $x_i = (D_i + z_i^2)/y_i$. The inequality to be proved is equivalent to $1/D_1 + 1/D_2 \geqslant 8/D$, or, since $D > 0$ (see below), to

$$(2) \qquad (D_1 + D_2)D \geqslant 8D_1D_2 .$$

Next we write D in terms of D_i, z_i and y_i:

$$\begin{aligned} D &= (x_1 + x_2)(y_1 + y_2) - (z_1 + z_2)^2 \\ &= x_1y_1 + x_1y_2 + x_2y_1 + x_2y_2 - z_1^2 - 2z_1z_2 - z_2^2 \\ &= D_1 + D_2 + x_1y_2 + x_2y_1 - 2z_1z_2 \\ &= D_1 + D_2 + \frac{D_1 + z_1^2}{y_1}y_2 + \frac{D_2 + z_2^2}{y_2}y_1 - 2z_1z_2 \\ &= D_1 + D_2 + \frac{D_1y_2}{y_1} + \frac{D_2y_1}{y_2} + \left(\frac{z_1}{y_1} - \frac{z_2}{y_2}\right)^2 y_1y_2 > 0 . \end{aligned}$$

Substituting this for D into (2), we obtain

$$(D_1 + D_2)^2 + (D_1 + D_2)\left(\frac{D_1y_2}{y_1} + \frac{D_2y_1}{y_2}\right) + (D_1 + D_2)\left(\frac{z_1}{y_1} - \frac{z_2}{y_2}\right)^2 y_1y_2 \geqslant 8D_1D_2 ,$$

and after subtracting $4D_1D_2$ from both sides, we find that the desired inequality is equivalent to

$$(3) \qquad (D_1 - D_2)^2 + (D_1 + D_2)\left[\frac{D_1y_2}{y_1} + \frac{D_2y_1}{y_2}\right] + (D_1 + D_2)\left(\frac{z_1}{y_1} - \frac{z_2}{y_2}\right)^2 \geqslant 4D_1D_2 .$$

Now (3) holds because the first and third terms are $\geqslant 0$, and the factors in the middle term, by the arithmetic-geometric mean inequality† $r + s \geqslant 2\sqrt{rs}$, satisfy

$$(4) \qquad \begin{aligned} D_1 + D_2 &\geqslant 2\sqrt{D_1D_2} , \\ \frac{D_1y_2}{y_1} + \frac{D_2y_1}{y_2} &\geqslant 2\sqrt{\frac{D_1D_2y_1y_2}{y_1y_2}} = 2\sqrt{D_1D_2} . \end{aligned}$$

Thus the middle term in (3) is $\geqslant 4D_1D_2$. Equality holds in (3) if and only if $D_1 = D_2$, $D_1y_2/y_1 = D_2y_1/y_2$, $z_1/y_1 = z_2/y_2$. The first two of these equations imply $y_1 = y_2$ which, with the third, gives $z_1 = z_2$. But these,

†See p. 196

with $D_1 = D_2$, yield $x_1 = x_2$. So equality holds if and only if $x_1 = x_2$, $y_1 = y_2$, $z_1 = z_2$.

Note. The purpose of this note is to generalize this problem. The quantities D_1, D_2 and D defined in (1) are determinants of the matrices

$$M_1 = \begin{pmatrix} x_1 & z_1 \\ z_1 & y_1 \end{pmatrix}, \quad M_2 = \begin{pmatrix} x_2 & z_2 \\ z_2 & y_2 \end{pmatrix} \quad \text{and} \quad M = M_1 + M_2$$

respectively, and the inequality to be proved may be written in the form

$$(5) \qquad \frac{8}{\det(M_1 + M_2)} \leqslant \frac{1}{\det M_1} + \frac{1}{\det M_2}\,.$$

An $n \times n$ matrix is *symmetric* if $a_{ij} = a_{ji}$ for all i, j. It is *positive definite* if the associated quadratic form $\Sigma a_{ij}u_iu_j$ is positive except at the origin.

We say that a function f of a variable y is *convex* on the interval $a \leqslant y \leqslant b$ if, for all y_1, y_2 in that interval

$$(6) \qquad f((1-t)y_1 + ty_2) \leqslant (1-t)f(y_1) + tf(y_2)\,.$$

For twice differentiable functions, convextiy in (a, b) is equivalent to $f''(y) \geqslant 0$ for $a \leqslant y \leqslant b$.

We can now state the generalization of the problem and sketch its proof.

THEOREM. *Inequality* (5) *holds for all positive definite matrices* M_1 *and* M_2 *of any order* n.

SKETCH OF PROOF: 1. If $p(t)$ is a polynomial with only real zeros, and if $p(t) > 0$ on $[a, b]$, then $1/p(t)$ is convex on $[a, b]$. (This can be proved by induction on the degree of $p(t)$.)

2. The function $\det[(1-t)M_1 + tM_2]$, where M_1 and M_2 are positive definite symmetric matrices, is a polynomial of degree n all of whose zeros are real, and it is positive for $0 \leqslant t \leqslant 1$. (This is an important property of symmetric positive definite matrices and can be proved by methods of linear algebra.)

3. The convexity of $1/\det[(1-t)M_1 + tM_2]$ implies inequality (5). (This follows from the convexity criterion (6) with $t = 1/2$ and from the fact that the determinant of a product M_1M_2 of two matrices is the product $\det M_1 \cdot \det M_2$ of their determinants.)

Twelfth International Olympiad, 1970

1970/1. Fig. 36a shows $\triangle ABC$, its incircle with center I and radius r, and the escribed circle (excircle) lying in $\angle ACB$ with center E and radius q. U and V are the points where the in- and ex-circles are tangent to AB. The measures of $\angle CAB$ and $\angle ABC$ are denoted by α and β. We shall equate two expressions for the length $AB = c$: $AU + BU = c$ and $AV + BV = c$. Now

$$(1) \quad AU = r\cot\frac{\alpha}{2}, \quad BU = r\cot\frac{\beta}{2}, \quad \text{so} \quad c = r\left(\cot\frac{\alpha}{2} + \cot\frac{\beta}{2}\right).$$

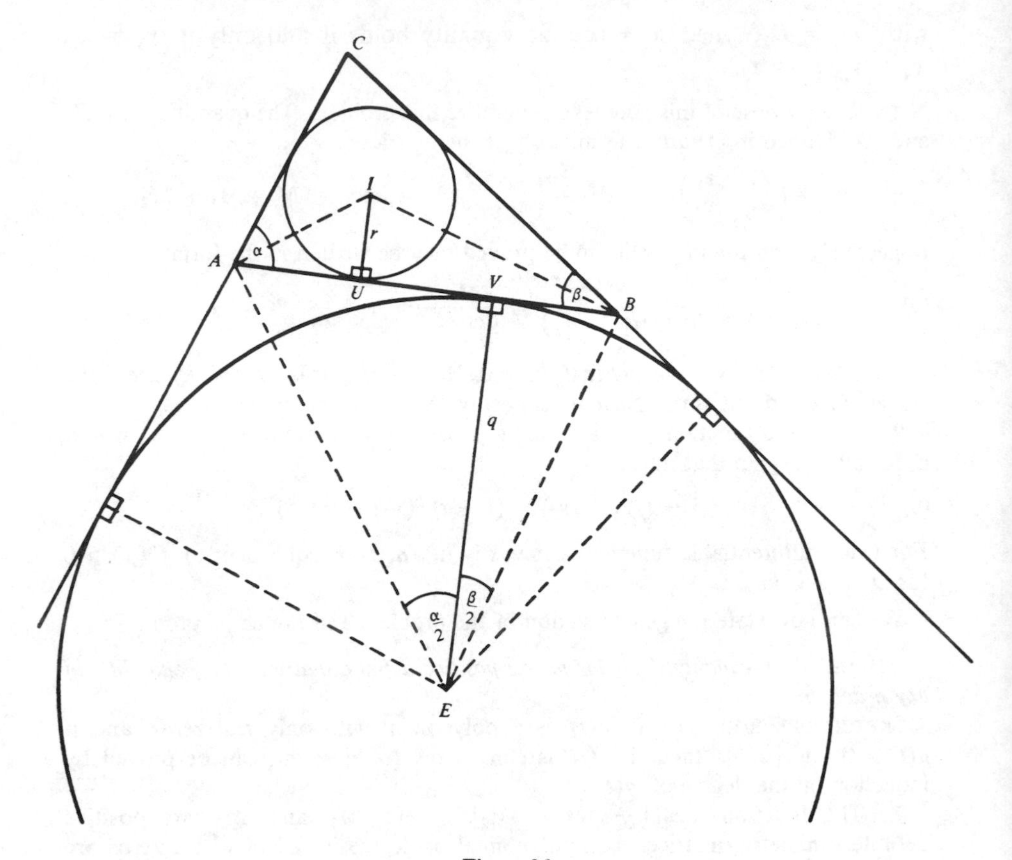

Figure 36a

Since the radii from E perpendicular to AB and BC form an angle whose sides are perpendicular to those of $\angle ABC$, $\angle BEV$ also has measure $\beta/2$. Similarly $\angle AEV = \alpha/2$. Therefore

(2) $AV = q \tan\dfrac{\alpha}{2}$, $BV = q\tan\dfrac{\beta}{2}$, so $c = q\left(\tan\dfrac{\alpha}{2} + \tan\dfrac{\beta}{2}\right)$.

Thus $r(\cot(\alpha/2) + \cot(\beta/2)) = q(\tan(\alpha/2) + \tan(\beta/2))$, and

$$\frac{r}{q} = \frac{\tan(\alpha/2) + \tan(\beta/2)}{\cot(\alpha/2) + \cot(\beta/2)} .$$

If we multiply the numerator and denominator of the right side by $\tan(\alpha/2)\tan(\beta/2)$, then divide by the common factor $\tan(\alpha/2) + \tan(\beta/2)$, we obtain

(3) $$\frac{r}{q} = \tan\frac{\alpha}{2}\tan\frac{\beta}{2} .$$

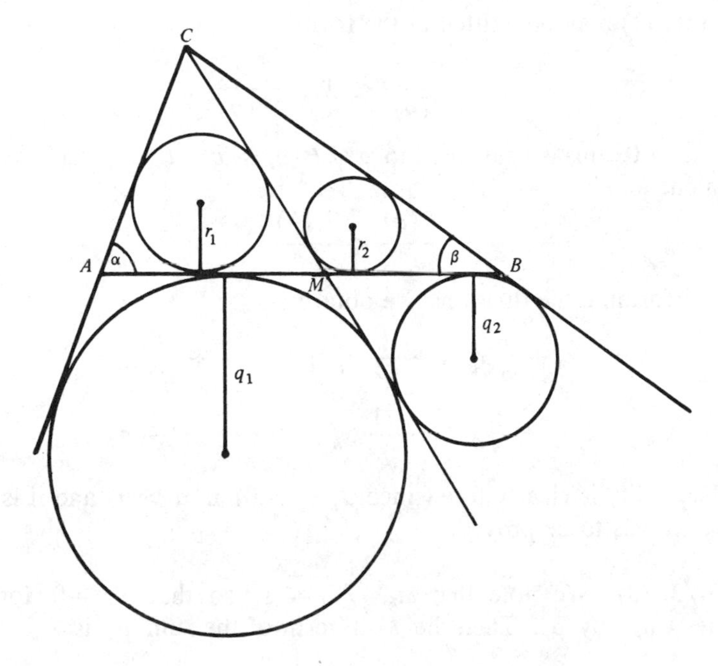

Figure 36b

Next we apply this result to the adjacent triangles $\triangle AMC$ and $\triangle MBC$, see Figure 36b. This yields

$$(4)\qquad \frac{r_1}{q_1} = \tan\frac{\alpha}{2}\tan\frac{\angle AMC}{2}\,, \qquad \frac{r_2}{q_2} = \tan\frac{\angle CMB}{2}\tan\frac{\beta}{2}\,.$$

Now $\angle AMC + \angle CMB = 180°$, so $\frac{1}{2}\angle CMB = 90° - \frac{1}{2}\angle AMC$, and $\tan\frac{1}{2}\angle CMB = \cot\frac{1}{2}\angle AMC$. Therefore

$$\frac{r_1}{q_1}\times\frac{r_2}{q_2} = \left(\tan\frac{\alpha}{2}\tan\frac{\angle AMC}{2}\right)\left(\cot\frac{\angle AMC}{2}\tan\frac{\beta}{2}\right)$$
$$= \tan\frac{\alpha}{2}\tan\frac{\beta}{2} = \frac{r}{q}\,.$$

1970/2. Let $P(t)$ be the polynomial

$$P(t) = x_n t^n + x_{n-1}t^{n-1} + \ldots + x_1 t + x_0\,.$$

Then

$$A_n = P(a),\ B_n = P(b),\ A_{n-1} = P(a) - x_n a^n,\ \ B_{n-1} = P(b) - x_n b^n\,.$$

In this notation, the assertion to be proved is

$$(1)\qquad \frac{P(a) - x_n a^n}{P(a)} < \frac{P(b) - x_n b^n}{P(b)} \qquad \text{if and only if } a > b\,.$$

Inequality (1) may be written in the form

$$1 - \frac{x_n a^n}{P(a)} < 1 - \frac{x_n b^n}{P(b)} ,$$

Since $x_n > 0$, this is equivalent to $a^n/P(a) > b^n/P(b)$ which, in turn, is equivalent to

$$\frac{P(a)}{a^n} < \frac{P(b)}{b^n} .$$

After performing the divisions, we obtain

$$x_n + \frac{1}{a} x_{n-1} + \frac{1}{a^2} x_{n-2} + \cdots + \frac{1}{a^n} x_0$$

$$< x_n + \frac{1}{b} x_{n-1} + \frac{1}{b^2} x_{n-2} + \cdots + \frac{1}{b^n} x_0 .$$

This inequality is clearly true (since $x_{n-1} \neq 0$) if $a > b$, and it is false if $a \leqslant b$, as was to be proved.

1970/3. (a) We note that $a_{k-1}/a_k \leqslant 1$, so that $b_n \geqslant 0$ for all n. Denote $\sqrt{a_k}$ by α_k. Then the k-th term of the sum b_n is

$$\left(1 - \frac{\alpha_{k-1}^2}{\alpha_k^2}\right)\frac{1}{\alpha_k} = \frac{\alpha_{k-1}^2}{\alpha_k}\left(\frac{1}{\alpha_{k-1}^2} - \frac{1}{\alpha_k^2}\right)$$

$$= \frac{\alpha_{k-1}^2}{\alpha_k}\left(\frac{1}{\alpha_{k-1}} + \frac{1}{\alpha_k}\right)\left(\frac{1}{\alpha_{k-1}} - \frac{1}{\alpha_k}\right)$$

$$= \frac{\alpha_{k-1}}{\alpha_k}\left(1 + \frac{\alpha_{k-1}}{\alpha_k}\right)\left(\frac{1}{\alpha_{k-1}} - \frac{1}{\alpha_k}\right)$$

$$\leqslant 2\left(\frac{1}{\alpha_{k-1}} - \frac{1}{\alpha_k}\right).$$

Adding these inequalities for $k = 1, 2, \ldots, n$, we observe that the right sides form a telescoping sum, and we get

$$0 \leqslant b_n \leqslant 2\left(\frac{1}{\alpha_0} - \frac{1}{\alpha_n}\right) = 2\left(\frac{1}{\sqrt{a_0}} - \frac{1}{\sqrt{a_n}}\right) = 2\left(1 - \frac{1}{\sqrt{a_n}}\right) < 2$$

for all n.

(b) Given $0 \leqslant c < 2$, we shall establish the existence of the appropriate a_i by constructing them as terms of a geometric series.

Set $1/\sqrt{a_k} = d^k$; then the k-th term of the sum b_n is

$$\left(1 - \frac{d^{-2(k-1)}}{d^{-2k}}\right)d^k = (1 - d^2)d^k .$$

Hence

$$b_n = \sum_{k=1}^{n} (1 - d^2)d^k = (1 - d^2) \sum_{k=1}^{n} d^k = (1 - d^2)\,\frac{d - d^{n+1}}{1 - d}$$

$$= d(1 + d)(1 - d^n)\ .$$

We must pick d between 0 and 1 so that $b_n = d(1 + d)(1 - d^n) > c$ for large enough n. We certainly need to have $d(1 + d) > c$. This can be achieved for any $c < 2$, since $d(1 + d)$ approaches 2 as d tends to 1. (In fact the reader can easily verify that $d(1 + d) > c$ if $d = \sqrt{c/2}$.) Now since $d < 1$, we see that $1 - d^n$ is as close to 1 as we please for all sufficiently large n. In particular, $1 - d^n > c/d(1 + d)$; that is, $d(1 + d)\,(1 - d^n) > c$ for all sufficiently large n. (We suggest that the reader find a number N depending on the given c such that $b_n > c$ for all $n > N$.)

1970/4 First solution. We shall show that there are no integers, positive or negative, having the prescribed property, even if instead of requiring equality of the products in each subset, we merely require them to be congruent modulo 7 (i.e., to have the same remainder when divided by 7).

Suppose the set $S = \{n, n + 1, n + 2, n + 3, n + 4, n + 5\}$ is partitioned into subsets S_1, S_2 such that the products a_1 and a_2 of the numbers of S_1 and S_2, respectively, are congruent mod 7:

$$a_1 \equiv a_2 (\text{mod } 7)\ . \qquad (1)$$

We claim that no number of S is divisible by 7; for, at most one of six consecutive numbers is divisible by 7. One of the subsets would contain it, the other not, so (1) could not hold. Therefore, the numbers in S are congruent to 1, 2, 3, 4, 5, 6 mod 7.

We shall use the following two facts about these residues mod 7.

(i) Their product satisfies $6! \equiv -1 (\text{mod } 7)$.

(ii) The congruence $x^2 \equiv -1 (\text{mod } 7)$ has no solution.

Both (i) and (ii) can be verified by direct calculation:

(i) $1 \cdot 2 \cdot 3 \cdot 4 \cdot 5 \cdot 6 \equiv 1 \cdot 2 \cdot 3(-3)(-2)(-1) \equiv -36 \equiv -1 (\text{mod } 7)$,

(ii) $1^2 \equiv 1,\ 2^2 \equiv 4,\ 3^2 \equiv 4^2 \equiv 2,\ 5^2 \equiv 4,\ 6^2 \equiv 1$.

Now since $a_1 \equiv a_2 (\text{mod } 7)$, the product of all the numbers in S, mod 7, is

$$a_1 \cdot a_2 \equiv a_1^2 (\text{mod } 7)\ .$$

On the other hand, the product of all these residues is $6! \equiv -1$; so we have

$$a_1^2 \equiv -1 (\text{mod } 7)\ ,$$

contradicting (ii). We conclude that S cannot be partitioned as required, for any integer n.

Remark. Relations (i) and (ii) are not peculiar to the prime 7. They can be generalized as follows:

(i)′ WILSON'S THEOREM. For any prime p,

$$(p-1)! \equiv -1 \pmod p .$$

(ii)′ For any prime p of the form $p = 4m + 3$,

$$x^2 \equiv -1 \pmod p \text{ has no solution} .$$

PROOF of (i)′: Observe that for each $y \not\equiv 0 \pmod p$, the numbers y, $2y$, $3y, \ldots, (p-1)y$ form a complete set of residues $(\operatorname{mod} p)$. Hence to each number y there exists a unique residue $y' \pmod p$ such that $yy' \equiv 1 \pmod p$; y' is called the inverse of $y \pmod p$. We evaluate $(p-1)!$ by pairing each factor with its inverse mod p; the product of these factors is 1. The only remaining factors are those which are their own inverses, i.e. those satisfying $x^2 \equiv 1 \pmod p$. They are $x \equiv 1 \pmod p$ and $x \equiv -1 \pmod p$. So the whole product is

$$(p-1)! \equiv 1(-1) \equiv -1 \pmod p$$

as Wilson's theorem asserts.

PROOF of (ii)′: Here we need Fermat's theorem, which states that for $y \not\equiv 0 \pmod p$, where p is a prime,

$$y^{p-1} \equiv 1 \pmod p .$$

This can be proved in many ways. For example, form the product

$$y(2y)(3y) \ldots (p-1)y = y^{p-1}(p-1)! .$$

Now the product of all the residues is congruent to $(p-1)!$, so

$$y^{p-1}(p-1)! \equiv (p-1)! \pmod p .$$

It follows that $y^{p-1} \equiv 1 \pmod p$.

To complete the proof of (ii)′, suppose $p = 4m + 3$ and that $x^2 \equiv -1 \pmod p$ had a solution. Then

$$x^{p-1} = x^{4m+2} = (x^2)^{(p-1)/2} \equiv (-1)^{(p-1)/2} \pmod p ,$$

and since $(p-1)/2 = 2m + 1$ is odd, we would obtain

$$x^{p-1} \equiv -1 \pmod p ,$$

contradicting Fermat's theorem. This completes the proof of (ii)′.

A number $a \not\equiv 0 \pmod p$ is called a *quadratic residue* mod p if the equation $x^2 \equiv a \pmod p$ has a solution. So (ii)′ says that -1 is not a quadratic residue of a prime p of the form $p = 4m + 3$. It can be shown that, for primes p of the form $p = 4m + 1$, -1 is a quadratic residue.

Second solution. Suppose the set S of six consecutive integers has been partitioned into non-empty sets S_1 and S_2 having the prescribed property. Then any prime factor p of an element of one of the sets is also a factor of an element of the other. However, if p divides two elements of the set S, say a and b, then $|a - b| = pk \leqslant 5$, so the only candidates for the prime p are 2, 3 and 5. A set of six consecutive integers contains at least one divisible by 5, and by the above argument, S must contain two such elements; these must be n and $n + 5$. The remaining elements $n + 1, n + 2, n + 3$ and $n + 4$ can have only 2 and 3 as prime factors, so they are of the form $2^\alpha 3^\beta$. Two of them are odd and two are even, so the odd ones are of the form 3^γ and 3^δ, where $\gamma, \delta > 0$. But they are consecutive odd numbers, so their difference is only 2, while the closest powers $3^\beta > 1$ have a difference of 3. Thus we see that there are no integers n with the prescribed property.

1970/5. We show first that each face angle at vertex D is a right angle. We are given that $\angle BDC$ is a right angle. In Figure 37, the plane CDH is perpendicular to plane ABC. Hence AB is perpendicular to the plane CDH, so $AB \perp DE$. We label the edges of the tetrahedron as shown in Figure 37:

$$a = BC, \quad b = CA, \quad c = AB, \quad p = AD, \quad q = BD, \quad r = CD .$$

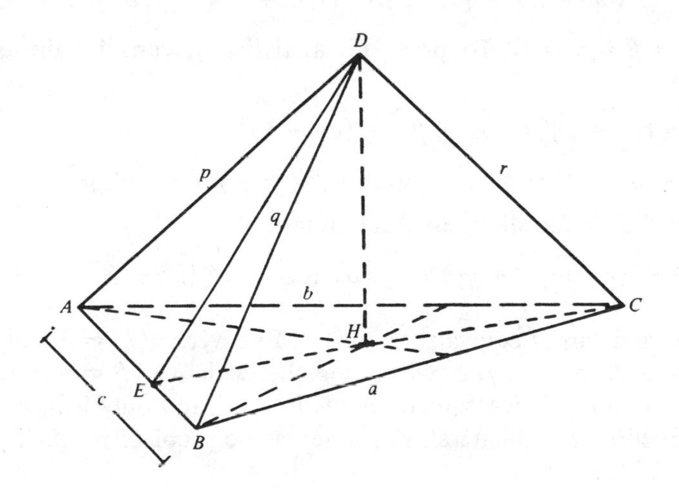

Figure 37

We find, by the Pythagorean theorem, that

(1) $$DE^2 + EB^2 = DB^2 = q^2,$$

(2) $$CE^2 + EB^2 = BC^2 = a^2 = q^2 + r^2 .$$

Subtracting (1) from (2) gives

$$CE^2 - DE^2 = r^2 \qquad \text{or} \qquad r^2 + DE^2 = CE^2 .$$

By the converse of the Pythagorean theorem, the last relation implies that

$CD \perp DE$. Since $CD \perp BD$ also, CD is perpendicular to plane ADB in which DE and DB lie. It follows that $CD \perp AD$, i.e. $\angle ADC = 90°$. By an entirely analogous argument, using the plane DHB, we can show that $\angle ADC = 90°$. Thus, for the three right-angled faces meeting at D, we have

$$q^2 + r^2 = a^2 , \quad p^2 + r^2 = b^2 , \quad p^2 + q^2 = c^2 ,$$

so that

$$(3) \qquad a^2 + b^2 + c^2 = 2(p^2 + q^2 + r^2) .$$

We shall show that

$$(4) \qquad (a + b + c)^2 \leqslant 3(a^2 + b^2 + c^2) ,$$

which, together with (3), implies the desired result

$$(a + b + c)^2 \leqslant 6(p^2 + q^2 + r^2) .$$

Inequality (4) is a special case of the Cauchy inequality†

$$(5) \qquad (\alpha a + \beta b + \gamma c)^2 \leqslant (\alpha^2 + \beta^2 + \gamma^2)(a^2 + b^2 + c^2)$$

with $\alpha = \beta = \gamma = 1$. To prove (5) analytically, consider the quadratic function of x

$$\begin{aligned} q(x) &= (ax + \alpha)^2 + (bx + \beta)^2 + (cx + \gamma)^2 \\ &= (a^2 + b^2 + c^2)x^2 + 2(\alpha a + \beta b + \gamma c)x + \alpha^2 + \beta^2 + \gamma^2 . \end{aligned}$$

Since $q(x) \geqslant 0$ for all x, its discriminant

$$[2(\alpha a + \beta b + \gamma c)]^2 - 4(a^2 + b^2 + c^2)(\alpha^2 + \beta^2 + \gamma^2)$$

is $\leqslant 0$, and this is equivalent to (5). Moreover, $q(x) = 0$ only if $x = -\alpha/a = -\beta/b = -\gamma/c$ which, together with $\alpha = \beta = \gamma = 1$, implies that $a = b = c$. Hence equality holds in (4) if and only if face ABC of the tetrahedron is equilateral. For a geometric proof of inequality (5), see Remark 2 below.

Remark 1. For readers familiar with vectors, we present another, perhaps more systematic proof of the key equation (3). Take H as the origin O, and denote the vectors from O to points A, B, C and D also by A, B, C and D; denote the scalar product of two vectors U and V by $U \cdot V$. Since $HA \perp BC$, we have

$$A \cdot (B - C) = 0 , \quad \text{i.e.} \quad A \cdot B = A \cdot C .$$

Similarly, since $HB \perp AC$, it follows that $B \cdot (A - C) = 0$, and hence $A \cdot B = B \cdot C$. Thus the scalar products $A \cdot B$, $A \cdot C$ and $B \cdot C$ are all equal:

$$A \cdot B = A \cdot C = B \cdot C .$$

†See p. 200.

Moreover HD is perpendicular to the plane ABC, so

$$A \cdot D = B \cdot D = C \cdot D = 0 .$$

From the fact that $\angle BDC = 90°$, we obtain $(B - D) \cdot (C - D) = 0$, and hence

$$B \cdot C + D \cdot D = 0 .$$

Therefore

$$a^2 + b^2 + c^2 = (C - B) \cdot (C - B) + (A - C) \cdot (A - C) + (B - A) \cdot (B - A)$$
$$= 2(A \cdot A + B \cdot B + C \cdot C) - 2(B \cdot C + C \cdot A + A \cdot B) ,$$

while

$$p^2 + q^2 + r^2 = (A - D) \cdot (A - D) + (B - D) \cdot (B - D) + (C - D) \cdot (C - D)$$
$$= A \cdot A + B \cdot B + C \cdot C + 3D \cdot D .$$

Since $-(B \cdot C + C \cdot A + A \cdot B) = -3B \cdot C = 3D \cdot D$, we conclude that $a^2 + b^2 + c^2 = 2(p^2 + q^2 + r^2)$.

Remark 2. A geometric proof of (5) can be obtained by interpreting the triples (a, b, c) and (α, β, γ) as components of vectors u and v. Their dot product is

$$u \cdot v = \alpha a + \beta b + \gamma c = |u||v| \cos \theta ,$$

where θ is the angle between u and v. Squaring and using the inequality $\cos^2\theta \leqslant 1$, we get (5), with equality if and only if u and v are collinear. For $v = (1, 1, 1)$, collinearity of u and v means that $a = b = c$.

1970/6. We shall show by induction that if $A(n)$ is the maximum number of acute angled triangles which can be formed from n points and $T(n) = \binom{n}{3}$ is the total number of triangles, then the ratios $A(n)/T(n)$ form a non-increasing sequence. We shall show that $A(5)/T(5) = .7$. Hence for $n > 5$, and in particular for $n = 100$, $A(n)/T(n) \leqslant .7$.

To treat the case $n = 5$, we need information about $n = 4$.

n = 4

Case (a) One of the points, say P_4, is inside the triangle formed by the other three. The three angles at P_4 add up to 360°; and since each is less than 180°, at least two are greater than 90°. Hence in this case there are at least two obtuse triangles (i.e. at most two acute triangles).

Case (b) The four points from a convex quadrilateral. Since the sum of its interior angles is 360°, at least one of them is $\geqslant 90°$, yielding at least one non-acute triangle. In this case three out of four triangles may be acute; therefore $A(4)/T(4) = .75$.

n = 5

Case (a) The convex hull of five points P_1, P_2, P_3, P_4, P_5 (i.e. the smallest convex set containing them) is a triangle, say $P_1P_2P_3$. Then at least two of the angles $P_1P_4P_2$, $P_2P_4P_3$, $P_3P_4P_1$ are not acute; and similarly, at least two of the angles $P_1P_5P_2$, $P_2P_5P_3$, $P_3P_5P_1$ are not acute. Thus we have at least four non-acute triangles.

Case (b) The convex hull is a quadrilateral, say $P_1P_2P_3P_4$, and P_5 is inside; see Figure 38a. Then P_5 is inside two of the $\binom{4}{3} = 4$ triangles that can be formed by points P_1, P_2, P_3, P_4 (in triangles $P_1P_2P_3$ and $P_2P_3P_4$ of Fig. 38a). As we saw for $n = 4$, case (a), we get two pairs ($P_1P_3P_5$, $P_2P_3P_5$ and $P_2P_3P_5$, $P_2P_4P_5$ in the figure) of non-acute triangles with vertex P_5; but these pairs may have a common member ($P_2P_3P_5$ in Fig. 38a), so we can count on at least 3 non-acute triangles with vertex P_5. Moreover one of the angles P_1, P_2, P_3, P_4 must be $\geqslant 90°$, giving still another non-acute triangle.

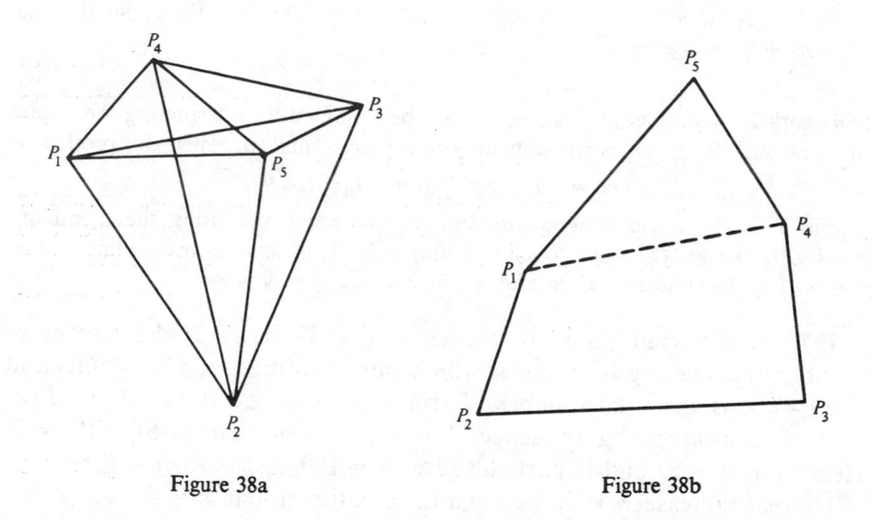

Figure 38a Figure 38b

Case (c) The points form a convex pentagon $P_1P_2P_3P_4P_5$. The sum of the interior angles is 540°, so at least two are non-acute, giving two non-acute triangles of the form $P_{i-1}P_iP_{i+1}$. If all three of the remaining interior angles are acute, two of them must be adjacent, say at P_2 and P_3, and by case (b) for $n = 4$ applied to quadrilateral $P_1P_2P_3P_4$, one of the angles $P_1P_4P_3$ or $P_4P_1P_2$ is non-acute, (see Fig. 38b), giving a total of three non-acute triangles.

Thus, in any configuration of five points, at least 3 triangles are non-acute. Since the total number of triangles is $\binom{5}{3} = 10$, at most 7 are acute, i.e. $A(5) \leqslant 7$. Figure 38b shows that equality can occur, so $A(5) = 7$, and $A(5)/T(5) = 7/10 = .7$.

Assume $A(n) \leqslant cT(n)$ for all integers from 5 up to n. We claim that then $A(n+1) \leqslant cT(n+1)$. To prove this, omit from the set S of $n+1$

points, in succession, the first point, the second point, ..., the $(n+1)$-st point. This yields $n+1$ sets of n points. Denote by B_k the number of acute triangles formed when the k-th point is omitted. By the induction hypothesis, $B_k \leqslant A(n) \leqslant cT(n)$ for each k. To count the total number B of acute triangles in the original set S, we form the sum $B_1 + B_2 + \ldots + B_{n+1}$. Each acute triangle is counted $(n+1) - 3 = n - 2$ times in this sum, since there are $n-2$ ways in which a non-vertex of each such triangle can be deleted from S. Hence

$$B = \frac{1}{n-2}(B_1 + B_2 + \ldots + B_{n+1}) \leqslant c\frac{n+1}{n-2}T(n) = cT(n+1) \,.$$

(Here we have used the fact that $[(n+1)/(n-2)]\binom{n}{3} = \binom{n+1}{3}$.) Since this inequality holds for any set of $n+1$ points, it follows that $A(n+1) \leqslant cT(n+1)$, and the induction is complete.

Remark. Our solution demonstrated that the maximum $R_n = A(n)/T(n)$ of the ratio of the number of acute-angled triangles to the total number of triangles is .75 for $n = 4$, .70 for $n = 5$, and does not increase, i.e.

$$R_4 = .75 \;, R_5 = .7 \geqslant R_6 \geqslant R_7 \geqslant \ldots \geqslant R_n \geqslant \ldots \,.$$

We know that such an infinite sequence has a limit, and it would be interesting to find out what this limit is. One way of getting a bit closer to the answer is to investigate the situation for very large n, at least in some special configurations. One such configuration consists of n points uniformly distributed along the circumference of a circle. This yields a lower bound of .25 for the R_n.

Thirteenth International Olympiad, 1971

1971/1. Denote the given expression by A_n. For $n = 3$, we give two solutions.

(i) Because of symmetry we can assume that $a_1 \geqslant a_2 \geqslant a_3$. Then

$$\begin{aligned} A_3 &= (a_1 - a_2)(a_1 - a_3) + (a_2 - a_1)(a_2 - a_3) + (a_3 - a_1)(a_3 - a_2) \\ &= (a_1 - a_2)\big[(a_1 - a_3) - (a_2 - a_3)\big] + (a_3 - a_1)(a_3 - a_2) \\ &= (a_1 - a_2)^2 + (a_3 - a_1)(a_3 - a_2) \\ &\geqslant 0. \end{aligned}$$

(ii) We also have

$$\begin{aligned} A_3 &= (a_1 - a_2)(a_1 - a_3) + (a_2 - a_1)(a_2 - a_3) + (a_3 - a_1)(a_3 - a_2) \\ &= a_1^2 - a_1a_3 - a_1a_2 + a_2^2 - a_2a_3 + a_3^2 \\ &= \left(\tfrac{1}{2}a_1^2 - a_1a_2 + \tfrac{1}{2}a_2^2\right) + \left(\tfrac{1}{2}a_2^2 - a_2a_3 + \tfrac{1}{2}a_3^2\right) + \left(\tfrac{1}{2}a_3^2 - a_1a_3 + \tfrac{1}{2}a_1^2\right) \\ &= \tfrac{1}{2}(a_1 - a_2)^2 + \tfrac{1}{2}(a_2 - a_3^2) + \tfrac{1}{2}(a_3 - a_1)^2 \;\; \geqslant \;\; 0 \,. \end{aligned}$$

To prove that $A_5 \geqslant 0$, we use the idea of proof (i) above that $A_3 \geqslant 0$. We can assume that $a_1 \geqslant a_2 \geqslant a_3 \geqslant a_4 \geqslant a_5$. From the first two terms of A_5 we take out the common factor $a_1 - a_2 \geqslant 0$, and write their sum as

$$(a_1 - a_2)[(a_1 - a_3)(a_1 - a_4)(a_1 - a_5) - (a_2 - a_3)(a_2 - a_4)(a_2 - a_5)] .$$

The difference in brackets is non-negative, since each factor in the first term is at least as great as the corresponding factor in the second term.

Similarly, we rewrite the sum of the last two terms of A_5 as

$$\begin{aligned}(a_4 - a_5)&[(a_4 - a_1)(a_4 - a_2)(a_4 - a_3) + (a_1 - a_5)(a_2 - a_5)(a_3 - a_5)]\\ &= (a_4 - a_5)[(a_1 - a_5)(a_2 - a_5)(a_3 - a_5) - (a_1 - a_4)(a_2 - a_4)(a_3 - a_4)] ,\end{aligned}$$

and observe that this expression is non-negative.

In the third term of A_5,

$$(a_3 - a_1)(a_3 - a_2)(a_3 - a_4)(a_3 - a_5) ,$$

the first two factors are non-negative, the last two non-positive, and the term is therefore non-negative.

In the case $n = 3$, we proved that $A_3 \geqslant 0$ in (ii) by writing A_3 as a sum of squares. This is a natural attack on the problem, since A_3 is a homogeneous quadratic function of the three variables a_1, a_2, a_3, and by one of the basic theorems of linear algebra, a quadratic form is non-negative if and only if it can be written as a sum of squares of linear forms.

In a note following this solution we show that it is not possible to treat the case $n = 5$ in this way, because the quadratic form A_5 cannot be written as a sum of squares of quadratic forms. This is related to a famous problem posed by Hilbert.

To show that the given relation is false for all other $n > 2$, it suffices to show that it is violated for a particular set of a_i. Now when n is even, A_n is a homogeneous form of odd degree, since each term consists of a product of $n - 1$ a_i's. Suppose the numbers a_i are replaced by their negatives; then A_n changes sign:

$$A_n(-a_1, -a_2, \ldots, -a_n) = -A_n(a_1, \ldots, a_n) .$$

This shows that A_n is not always $\geqslant 0$ for n even.

When n is odd and greater than 5, we assign the following values to the a_i:

$$a_1 = a_2 = \ldots = a_{n-4} = 0 , \qquad a_{n-3} = 1 , \qquad a_{n-2} = a_{n-1} = a_n = 2 .$$

Then all terms of A_n except the $(n - 3)$rd term vanish, and

$$A_n = (1)^{n-4}(1 - 2)^3 = (-1)^3 < 0 .$$

Note on the Representation of Forms as Sums of Squares. We shall show now that the fourth order form (i.e. the homogeneous polynomial of degree 4)

(1) $$A_5 = \prod_{j\neq 1}(a_1 - a_j) + \prod_{j\neq 2}(a_2 - a_j) + \prod_{j\neq 3}(a_3 - a_j) + \prod_{j\neq 4}(a_4 - a_j) + \prod_{j\neq 5}(a_5 - a_j)$$

cannot be written as a sum of squares of quadratic forms in $a_1, \ldots, a_5$. For suppose on the contrary that

(2) $$A_5 = \sum_{j=1}^{5} Q_j^2 ,$$

where Q_j is a quadratic form in some or all of the variables $a_1, \ldots, a_5$. Then whenever $A_5 = 0$, each $Q_j = 0$. Now A_5 clearly vanishes if the a_i satisfy one of the equations resulting from

(3) $$a_1 = a_2 \quad \text{and} \quad a_3 = a_4 = a_5$$

when the indices 1, 2, 3, 4, 5 are permuted. We shall show below that a quadratic form in $a_1, a_2, \ldots, a_5$ which vanishes under conditions (3) and under all permutations of (3), vanishes identically. But A_5 does not vanish identically, so A_5 cannot be of the form (2).

It remains to prove the following

LEMMA. *Let Q be a quadratic form in $a_1, a_2, \ldots, a_5$ which is zero whenever the a_i satisfy* (3) *or some permutation of* (3). *Then $Q = 0$ for all values of the a_i.*

PROOF: We may write

$$Q(a_1, a_2, a_3, a_4, a_5) = \sum_{k,l=1}^{5} c_{kl} a_k a_l ,$$

where the coefficients c_{kl} are symmetric, i.e.

$$c_{kl} = c_{lk} .$$

By hypothesis, we have

(4) $$\begin{aligned} Q(1, 1, 0, 0, 0) &= c_{11} + c_{22} + 2c_{12} = 0 \\ Q(1, 0, 1, 0, 0) &= c_{11} + c_{33} + 2c_{13} = 0 \\ Q(0, 1, 1, 0, 0) &= c_{22} + c_{33} + 2c_{23} = 0 . \end{aligned}$$

Adding these three equations, we obtain

(5) $$2c_{11} + 2c_{22} + 2c_{33} + 2c_{12} + 2c_{13} + 2c_{23} = 0 .$$

Next we use the fact that

$$Q(1, 1, 1, 0, 0) = c_{11} + c_{22} + c_{33} + 2c_{12} + 2c_{13} + 2c_{23} = 0 .$$

Subtracting this from (5), we get

(6) $$c_{11} + c_{22} + c_{33} = 0 .$$

Since this equation also holds after any permutation of the subscripts, we have

(7) $$c_{11} + c_{22} + c_{44} = 0 .$$

A comparison of (6) and (7) yields

$$c_{33} = c_{44} .$$

Since this also holds for all permutations of the subscripts, we have

$$c_{11} = c_{22} = c_{33} = c_{44} = c_{55} .$$

Substituting this into (6), we find that all the c_{ii} are zero. Hence by (4), we have $c_{12} = 0$. By permuting the subscripts, we then get $c_{ij} = 0$ for all i, j. Thus Q is identically zero.

Our result shows that the quartic form A_5, although non-negative, cannot be written as a sum of squares of quadratic forms. Thus quartic forms behave differently from quadratic forms in this respect.

The general theory of these forms goes back to Hilbert, who showed that every nonnegative quartic form in three variables can be written as a sum of squares, but gave an example of a quartic form in four variables which cannot be so represented.

Our example A_5 depends on the five variables $a_1, \ldots, a_5$; but since only their differences enter, and since each such difference can be expressed in terms of the four quantities

$$d_1 = a_1 - a_2 , \qquad d_2 = a_2 - a_3 , \qquad d_3 = a_3 - a_4 , \qquad d_4 = a_4 - a_5$$

(for example, $a_2 - a_5 = d_2 + d_3 + d_4$, $a_4 - a_1 = -(d_1 + d_2 + d_3)$ etc.), we see that A_5 is essentially a quartic form in four variables.

In his famous list of 23 problems, Hilbert asked the following question: Can every nonnegative form of any degree and in any number of variables be written in the rational form

$$(\Sigma Q_j^2)/R^2 ,$$

where Q_j and R are forms? That this is indeed always possible was proved by E. Artin in 1928.

1971/2. Let A_1 be placed at the origin of a three-dimensional coordinate system, and denote the vectors to the vertices A_i again by A_i. Let D be the polyhedron obtained from P_1 by enlarging it in the ratio 2:1; that is, by replacing the vectors A_i to the vertices by $2A_i$. Clearly D contains P_1. We claim that D also contains all the translates $P_2, P_3, \ldots, P_9$. To see this, let X_i be the vector to a point in P_i, and let X_1 be the vector to the corresponding point in P_1. Then

$$X_i = A_i + X_1 = 2\left(\tfrac{1}{2}A_i + \tfrac{1}{2}X_1\right) .$$

Since A_i and X_1 are points of P_1, and since P_1 is convex, the midpoint

$\frac{1}{2}(A_i + X_1)$ of the segment A_iX_1 is also in P_1. Hence X_i is in D, which proves our claim.

Now the volume of D is 8 times the volume of P_1. Consequently, the 9 translates of P_1 contained in D cannot have disjoint interiors.

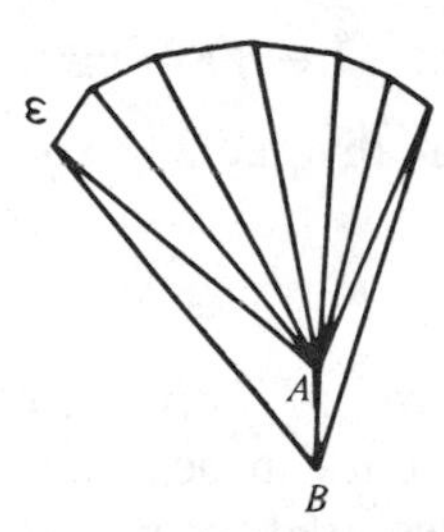

Figure 39

Remark. The theorem tells us that if we translate every corner of P_1 to some fixed point O, (i.e. if we move every vertex A_j along A_jO to O), then there is a ray through O in the interior of at least two of the nine resulting solid angles at O. A corresponding theorem in the plane for convex polygons of more than four vertices follows from the fact that the sum of the interior angles exceeds 360°. No such proof can be given for the three-dimensional theorem. The sum of the solid angles of a tetrahedron can be made arbitrarily small by taking two opposite edges short enough and assigning a fixed length to the other four. For example, take AB and CD of length ε, the other edges of length 1, and join 6 such congruent tetrahedra, so that AB is a common edge (see Fig. 39). The resulting convex polyhedron with 9 vertices has solid angles whose sum is as small as we want.

1971/3. We shall give a recipe for actually constructing an infinite set of integers of the form

$$a_i = 2^{k_i} - 3 \qquad i = 1, 2, \ldots,$$

each relatively prime to all the others.

Suppose we have n relatively prime members

$$a_1 = 2^{k_1} - 3, \qquad a_2 = 2^{k_2} - 3, \qquad \ldots, \qquad a_n = 2^{k_n} - 3.$$

[For example, $a_1 = 2^3 - 3 = 5$ and $a_2 = 2^4 - 3 = 13$ is such a set for $n = 2$. Here $k_1 = 3$ and $k_2 = 4$.] We form the product

$$(1) \qquad s = \prod_{i=1}^{n} a_i = (2^{k_1} - 3)(2^{k_2} - 3) \ldots (2^{k_n} - 3)$$

and observe that s, a product of odd numbers, is odd. [In our example, $s = 5 \cdot 13 = 65$.]

Now consider the set of $s+1$ numbers $2^0, 2^1, 2^2, \ldots, 2^s$. Since, when an integer is divided by s, there are only s possible remainders $0, 1, 2, \ldots, s-1$, at least two of these $s+1$ numbers, say 2^α and 2^β, $\alpha > \beta$, leave the same remainder on division by s; i.e. $2^\alpha \equiv 2^\beta \pmod{s}$, or

$$2^\beta(2^{\alpha-\beta}-1) = ms\ , \qquad m \text{ an integer}\ .$$

The odd number s does not divide 2^β, so it must divide $2^{\alpha-\beta}-1$; hence

$$2^{\alpha-\beta}-1 = ls\ , \qquad l \text{ an integer}\ .$$

Since $2^{\alpha-\beta}-1$ is divisible by s, and since s is odd, $2^{\alpha-\beta}-3$ is relatively prime to s. Therefore it can be adjoined as a new member a_{n+1} to the desired set. (It has the required form, and it is relatively prime to all the others, since it is relatively prime to their product s.) Repeated application of this construction leads to an infinite subset of relatively prime integers. [In our example, $2^{12}-2^0 = 4095 = 63 \cdot 65$; i.e., $\alpha = 12$, $\beta = 0$, $k_3 = 12 - 0 = 12$, $a_3 = 2^{12} - 3 = 4093$. The numbers 2^3-3, 2^4-3, $2^{12}-3$ are relatively prime.]

Remark. The key ingredient in the above solution is finding an exponent k such that $2^k - 1$ is divisible by s; then we argued that $2^k - 3$ has no factor in common with the odd number s. We established the existence of such a k by applying the pigeon hole principle. However, there is also a famous theorem which not only tells us that such an exponent exists, but gives it to us explicitly. Before applying it to our problem, we state and prove this theorem.

The *Euler function* $\phi(s)$ is defined to be the number of positive integers $\leqslant s$ and relatively prime to s.

Euler-Fermat Theorem. If y is relatively prime to s, then

$$y^{\phi(s)} \equiv 1 \pmod{s}\ . \tag{2}$$

PROOF: We recall our proof of Fermat's theorem (see the remark following the solution of 1970/4, p. 126) and modify it. Instead of forming the product $y \cdot 2y \cdot 3y \cdot \cdots \cdot (s-1)y$, we use only the $\phi(s)$ residues $r_1, r_2, \ldots, r_\phi$ relatively prime to s, and write

$$y \cdot r_1 y \cdot r_2 y \cdot \cdots \cdot r_\phi y = y^{\phi(s)} r_1 r_2 \cdots r_\phi \equiv r_1 r_2 \cdots r_\phi \pmod{s}\ .$$

It follows that $y^{\phi(s)} \equiv 1 \pmod{s}$.

Since s, defined by (1), is odd, 2 is relatively prime to s; so by (2), $2^{\phi(s)} - 1$ is divisible by s. It follows as before that then $2^{\phi(s)} - 3$ has no factor in common with s, and we may admit $a_{n+1} = 2^{\phi(s)} - 3$ as a new member of the desired set.

1971/4. The path does not cut edges AC and BD; see Fig. 40a. Cut the tetrahedron along edges BD, AC and DA and unfold triangles BCD

and ABD, so that they lie in the plane of $\triangle ABC$; then unfold $\triangle ACD$, attached to $\triangle BCD$ along CD, so that it also lies in the same plane. Now we have the configuration of Fig. 40b, with segment AD on the left marked A_1D_1 for clarity. The path we are to minimize is the broken line $TXYZT$.

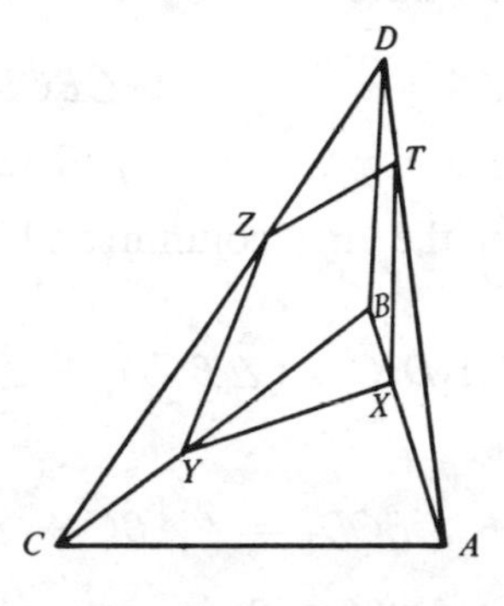

Figure 40a

Consider path XYZ; if it is not straight, i.e., if Y is not on the segment XZ, then this path can be shortened by moving the point Y to the intersection Y' of XZ and BC. This is always possible since acute triangles ABC and BCD_1 form a convex quadrilateral ABD_1C, so any segment from a point on AB to a point on CD_1 intersects the diagonal BC. The same argument holds for the path TXY. Thus we see that if path $TXYZT$ is not straight, it can always be shortened by removal of a kink in the path.

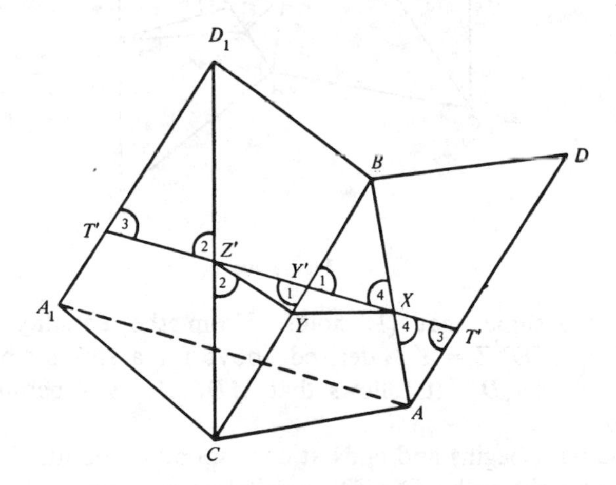

Figure 40b

(a) Suppose $T'X'Y'Z'T'$ is straight. Then the angles with the same labels in Fig. 40b are equal vertical angles. To see that this is true for the angles marked 3, observe that the tetrahedron could also be unfolded as in Fig. 40c. Then the above reasoning shows that for a minimal path, $X'T'Z'$ is a straight line and intersects AD_1 at T' forming the equal vertical angles 3. We now have

$$\angle ABC + \angle 1 + \angle 4 = \pi , \qquad \angle BCD_1 + \angle 1 + \angle 2 = \pi ,$$
$$\angle A_1D_1C + \angle 2 + \angle 3 = \pi , \qquad \angle DAB + \angle 3 + \angle 4 = \pi .$$

Adding the equations in the first column and subtracting those in the second, we obtain

$$\angle ABC + \angle A_1D_1C - (\angle BCD_1 + \angle DAB) = 0 ,$$

whence

$$\angle DAB + \angle BCD_1 = \angle ABC + \angle A_1D_1C . \tag{1}$$

Therefore, if (1) is violated, the path is not straight, hence not minimal. This completes the proof of part (a).

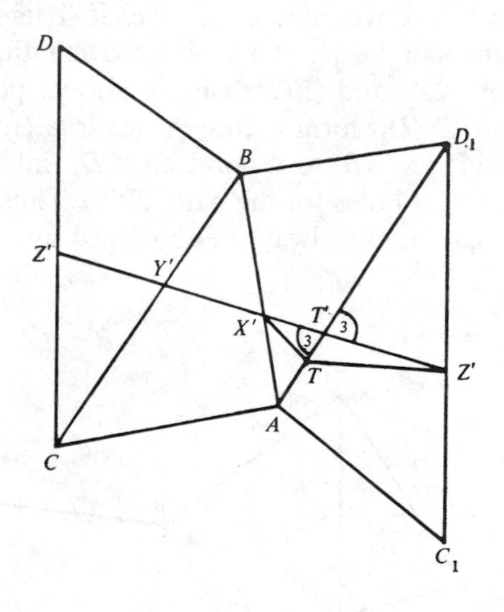

Figure 40c

(b) Now assume that (1) holds. From the equality $\angle AT'X' = \angle D_1T'Z' = \angle DT'Z = \angle 3$ derived above for a straight path, we conclude that $AD \| A_1D_1$. It follows that ADD_1A_1 is a parallelogram, see Figure 40b.

Since the path begins and ends at corresponding points T on opposite sides, any straight path $TXYZT$ parallel to segment AA_1 has minimal length. AA_1 is the base of isosceles triangle ACA_1 with vertex angle

$\angle A_1CD_1 + \angle D_1CB + \angle BCA = \alpha$, and therefore it has length $2AC\sin(\alpha/2)$.

1971/5. With the help of m unit vectors in the plane, we shall construct a set S consisting of 2^m points and having the required property. We shall see that certain easily met requirements have to be put on the m unit vectors $u_1, u_2, \ldots, u_m$.

Construction. Form the vectors

(1) $v = c_1u_1 + c_2u_2 + \ldots + c_mu_m$, where $|u_i| = 1$, $i = 1, 2, \ldots, m$,

and where each coefficient c_i may take one of the two values $1/2$ and $-1/2$; thus 2^m such vectors can be formed. Their endpoints form our set S, provided we choose the unit vectors $u_1, u_2, \ldots, u_m$ in the proper fashion, as we shall see.

Now consider a particular vector v of form (1), and define m vectors $v_1, v_2, \ldots, v_m$ by

(2) $v_i = c_1u_1 + c_2u_2 + \ldots + c_{i-1}u_{i-1} - c_iu_i + c_{i+1}u_{i+1} + \ldots + c_mu_m$.

In other words, v_i differs from v only in the i-th coefficient; if the i-th coefficient of v is $1/2$, that of v_i is $-1/2$, and vice versa. It follows that $v - v_i = \pm u_i$, $i = 1, 2, \ldots, m$, so $|v - v_i| = 1$; i.e., each v_i is one unit from v.

Restrictions on the m unit vectors u_i. To show that for each v in S, there are *exactly* m points in the set S at distance 1 from v, we must show that for fixed v,

(a) the m vectors v_i defined by (2) are distinct, and
(b) no other vector w in S has distance 1 from v.

Condition (a) means that $v_i - v_j \neq 0$ for $i \neq j$. Now by (2)

$$v_i - v_j = -2c_iu_i + 2c_ju_j .$$

Hence $v_i - v_j = 0$ only if $-c_iu_i + c_ju_j = 0$, i.e. if u_i and u_j are collinear. To rule this out, we require:

(3) no two of the unit vectors $u_1, u_2, \ldots, u_m$ are collinear.

[We can select the u_i inductively: having chosen $u_1, u_2, \ldots, u_k$ so that no two are collinear, we pick u_{k+1} different from the $2k$ vectors $\pm u_1$, $\pm u_2, \ldots, \pm u_k$.]

Condition (b) says that $|w - v| \neq 1$ whenever w in S is not one of the m vectors defined by (2). Since all the v_i differ from v in exactly one coefficient, this means that w differs from v in at least two coefficients. Thus condition (b) can be written:

(4) $$|w - v| = \left|\sum_1^m d_ku_k\right| = |d_1u_1 + d_2u_2 + \ldots + d_mu_m| \neq 1$$

for all d_k in the set $\{-1, 0, 1\}$ where at least two d_k are different from 0. Thus we can construct the desired set S if we can select m unit vectors satisfying conditions (3) and (4).

We now show by induction how to satisfy the requirements

$$|d_1u_1 + d_2u_2 + \ldots + d_mu_m| \begin{cases} \neq 0 \\ \neq 1 \end{cases}$$

for all d_k in the set $\{-1, 0, 1\}$, where at least two of the d_k are not zero. (The first of these requirements implies (3), and the second is equivalent to (4).)

For $m = 2$, take an arbitrary unit vector u_1; then pick the unit vector u_2 (a) so that it is not collinear with u_1, (b) so that its endpoint is not one of the four points where the unit circles with centers at the endpoints of u_1 and $-u_1$ intersect the unit circle around the origin. Now u_1, u_2 satisfy (3) and (4).

Suppose we have $m-1$ unit vectors $u_1, u_2, \ldots, u_{m-1}$ satisfying (3) and (4). If u_m is chosen so that it is different from any of the vectors $\Sigma_1^{m-1} d_iu_i$ with d_i in the set $\{-1, 0, 1\}$, then condition (3) will be satisfied by $u_1, u_2, \ldots, u_m$. Moreover, since the vectors $s = \Sigma_1^{m-1} d_iu_i$, $d_i \in \{-1, 0, 1\}$, do not vanish (by the induction hypothesis), the vector $s + u_m$ has length 1 only if it lies on an intersection of the unit circle centered at s with the unit circle around the origin. These unit circles have at most two intersections; so for each s, u_m can be picked so that $s + u_m$ avoids both these intersections, and also so that $s - u_m$ avoids them. In other words, conditions (3) and (4) place a finite number of restrictions on u_m, but we have a whole continuum of candidates for u_m among vectors ending at points on the unit circle around the origin. So u_m can be chosen so that (3) and (4) are satisfied.

1971/6. Consider the row sums $R_i = \Sigma_{j=1}^n a_{ij}$ $(i = 1, 2, \ldots, n,)$ and the column sums $C_j = \Sigma_{i=1}^n a_{ij}$ $(j = 1, 2, \ldots, n.)$ Denote by p the smallest of all these sums R_i and C_j:

$$p = \min_{i,j} \{R_i, C_j\} .$$

Case 1. Suppose $p \geqslant n/2$. Then the sum S of all elements of the matrix is at least $np \geqslant n(n/2)$, so $S \geqslant n^2/2$, and the assertion is true.

Case 2. Suppose $p < n/2$. By interchanging rows and columns, if necessary, and then reordering the rows, we can suppose that the elements of the first row have sum p; and by re-ordering columns, if necessary, we can assume that the first q elements in the first row are non-zero, while all other elements in the first row are zero. By hypothesis, the sum of the elements in each of the $n-q$ columns headed by 0 plus the sum p of the elements in the first row is $\geqslant n$. Therefore the sum of the elements in each of these columns is $\geqslant n - p$. Since there are $n - q$ such columns, the sum of all the elements in them is $\geqslant (n-p)(n-q)$. The sum of all elements in the first q columns is at least pq. Therefore the sum S of all the elements in the matrix satisfies

$$S \geqslant (n-p)(n-q) + pq = n^2 - np - nq + 2pq$$

$$= \tfrac{1}{2}n^2 + \tfrac{1}{2}(n-2p)(n-2q) .$$

Now $p < n/2$ by assumption, so $n - 2p > 0$. Moreover $q \leqslant n/2$, for if more than $n/2$ elements of the first row were positive integers, their sum p would be $\geqslant n/2$. Therefore $n - 2q \geqslant 0$, and $S \geqslant n^2/2$ in all cases.

Fourteenth International Olympiad, 1972

1972/1. If $S = \{a_1, a_2, \ldots, a_n\}$ is a set of n elements, the number of subsets of S (including the empty set and S itself) is 2^n. Perhaps the simplest of many ways of proving this is the following. If T is a subset of S, then for each i either $a_i \in T$ or $a_i \notin T$. Hence the number of possibilities for T is $2 \cdot 2 \ldots 2 = 2^n$. In particular, the number of subsets of the given set S of ten numbers is $2^{10} = 1024$.

The numbers in S are all $\leqslant 99$. Hence the sum of the numbers in any subset of S is $\leqslant 10 \cdot 99 = 990$.† So there are fewer possible sums than there are subsets. It follows from the pigeon-hole principle‡ that at least two different subsets, say S_1 and S_2, must have the same sum. If S_1 and S_2 are disjoint, the problem is solved. If not, we remove all elements common to S_1 and S_2 from both, obtaining non-intersecting subsets S_1' and S_2' of S. The sum of the numbers in S_1' is equal to the sum of the numbers in S_2'.

1972/2. Let $ABCD$ be an inscribable quadrilateral, and let angle A be its smallest angle. (If there are several smallest angles, $ABCD$ is an isosceles trapezoid and can be divided into n inscribable quadrilaterals merely by $n - 1$ segments parallel to the parallel bases.) Let P be a point inside the quadrilateral, and draw $PE \| AB$, $PF \| AD$. If P is close enough to A then E will lie between B and C, and F between C and D, as shown in Fig. 41. Also, draw PG so that $\angle PGD = \angle D$ and PH so that $\angle PHB = \angle B$. Since $\angle B > \angle A$, H will lie between A and B if P is close enough to A, and similarly G will lie between A and D.

†In fact, since S has distinct elements, the largest possible sum is only $99 + 98 + 97 + 96 + 95 + 94 + 93 + 92 + 91 + 90 = 945$.

‡See p. 201.

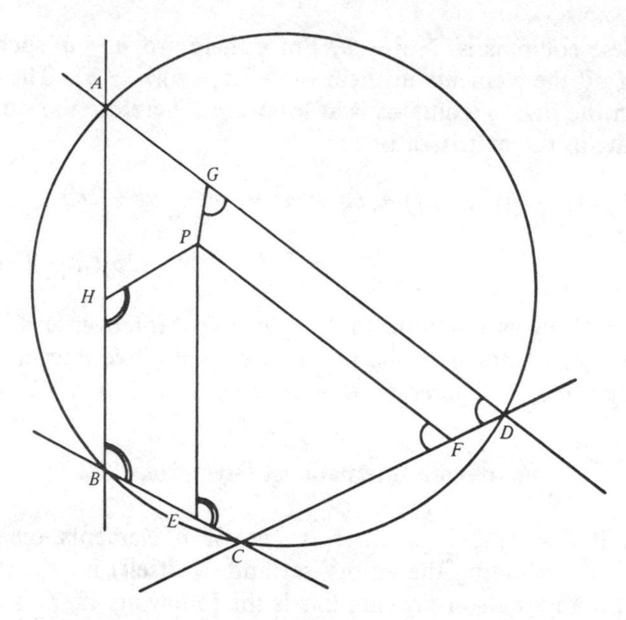

Figure 41

The quadrilateral $PECF$ is cyclic because it has the same angles as $ABCD$. $AHPG$ is cyclic because $\angle AHP + \angle AGP = (180° - \angle B) + (180° - \angle D) = 180°$. $PHBE$ and $PFDG$ are isosceles trapezoids, hence inscribable. We have subdivided $ABCD$ into 4 cyclic quadrilaterals. To subdivide into more, just draw lines parallel to the bases in one of the isosceles trapezoids.

1972/3 First solution. We observe that if $n = 0$, the given expression reduces to the binomial coefficient $\binom{2m}{m}$, which is an integer. It is not surprising that the given expression

$$f(m, n) = \frac{(2m)!\,(2n)!}{m!n!\,(m+n)!} \tag{1}$$

behaves somewhat like binomial coefficients in that it satisfies a recursion relation. In particular, the reader can verify directly that

$$f(m, n) = 4f(m, n-1) - f(m+1, n-1)\ . \tag{2}$$

Using this recurrence, we can easily prove by induction on n that $f(m, n)$ is an integer. As already remarked, $f(m, 0) = \binom{2m}{m}$ is an integer. If $n \geqslant 1$, and if it has already been shown that $f(m, n-1)$ is an integer for all m, then (2) implies that $f(m, n)$ is an integer. This completes the induction.

Second solution (due to Eric Lander). To show that the given expression is an integer, we show that its denominator

$$D = m!n!\,(m+n)!$$

divides its numerator

$$N = (2m)!\,(2n)!\ .$$

Denote by p^a the highest power of the prime p which divides D, and by p^b the highest power of p which divides N. Then D divides N if, for every prime p, $a \leqslant b$.

Now the greatest exponent α such that p^α divides $k!$ is given by

$$\alpha = \sum_{j=1}^{\infty} \left[\frac{k}{p^j} \right],$$

where $[x] =$ integer part of x. To see this, we observe that $[k/p]$ counts the number of integers among $1, 2, 3, \ldots, k$ which are divisible by p, each of which contributes 1 to the number of factors p in $k!$; $[k/p^2]$ counts the number of integers $\leqslant k$ divisible by p^2, hence contributing an additional factor p to the factorization of $k!$, etc.

Suppose that in the factorization of D and N, the prime p occurs a and b times respectively. Then

$$\text{(3)}\qquad \begin{aligned} a &= \sum_{j=1}^{\infty} \left[\frac{m}{p^j} \right] + \sum_{j=1}^{\infty} \left[\frac{n}{p^j} \right] + \sum_{j=1}^{\infty} \left[\frac{m+n}{p^j} \right] \\ &= \sum_{j=1}^{\infty} \left\{ \left[\frac{m}{p^j} \right] + \left[\frac{n}{p^j} \right] + \left[\frac{m+n}{p^j} \right] \right\}, \end{aligned}$$

and

$$\text{(4)}\qquad b = \sum_{j=1}^{\infty} \left[\frac{2m}{p^j} \right] + \sum_{j=1}^{\infty} \left[\frac{2n}{p^j} \right] = \sum_{j=1}^{\infty} \left\{ \left[\frac{2m}{p^j} \right] + \left[\frac{2n}{p^j} \right] \right\}.$$

To prove that $a \leqslant b$, it suffices to show that each term in the sum (3) is less than or equal to the corresponding term in the sum (4), i.e.

$$\text{(5)}\qquad \left[\frac{m}{p^j} \right] + \left[\frac{n}{p^j} \right] + \left[\frac{m+n}{p^j} \right] \leqslant \left[\frac{2m}{p^j} \right] + \left[\frac{2n}{p^j} \right].$$

We shall prove the inequality

$$\text{(6)}\qquad [r] + [s] + [r+s] \leqslant [2r] + [2s]$$

for any non-negative real r and s and deduce (5) by setting $r = m/p^j$, $s = n/p^j$.

It suffices to prove (6) for $0 \leqslant r < 1$ and $0 \leqslant s < 1$, for if either r or s is increased by 1, both sides of (6) increase by 2.

When $0 \leqslant r < 1$ and $0 \leqslant s < 1$, we have $[r] = [s] = 0$, and hence we must show that $[r + s] \leqslant [2r] + [2s]$ in this range. This is clear if $r + s < 1$. If $r + s \geqslant 1$, then either $r \geqslant 1/2$ or $s \geqslant 1/2$. In this case $[r + s] = 1$, while $[2r] + [2s] \geqslant 1$.

1972/4 First solution (due to Paul Zeitz). Each inequality is of the form

$$(1) \qquad (x_i^2 - x_{i+2}x_{i+4})(x_{i+1}^2 - x_{i+2}x_{i+4}) \leqslant 0\,,$$

where the indices are read modulo 5, i.e. $x_{j+5} = x_j$. If we multiply out each expression on the left and then add all the inequalities, we find that all $\binom{5}{4} = 10$ terms of the form $x_i^2x_j^2$ $(i \neq j)$ appear in the sum, as well as ten "cross terms", five of the form $-x_i^2x_{i+1}x_{i+3}$ and five of the form $-x_i^2x_{i+2}x_{i+4}$. This suggests a sum of squares of the form $\frac{1}{2}(y_1^2 + y_2^2 + \dots + y_{10}^2)$, where to each "cross term", we associate a y; for example, to the cross term $x_2^2x_3x_5$ we associate $y_1 = x_2x_3 - x_2x_5$, knowing that the terms $x_2^2x_3^2$, $x_2^2x_5^2$ appearing in y_1^2 also appear in our sum. Thus we arrive at the following representation of the sum of the given inequalities:

$$0 \geqslant \sum_{i=1}^{5} (x_i^2 - x_{i+2}x_{i+4})(x_{i+1}^2 - x_{i+2}x_{i+4})$$

$$= \frac{1}{2}\sum_{i=1}^{5}\left\{(x_ix_{i+1} - x_ix_{i+3})^2 + (x_{i-1}x_{i+1} - x_{i-1}x_{i+3})^2\right\}$$

Since this sum of squares cannot be negative. we conclude that it is zero, which means that each term vanishes. This implies that $x_1 = x_2 = x_3 = x_4 = x_5$. Every set of five equal positive numbers is a solution of the given system of inequalities.

Second solution. Each inequality states that the two factors in (1) have opposite signs, or that at least one vanishes. Thus either

$$x_i^2 \geqslant x_{i+2}x_{i-1} \qquad \text{and} \qquad x_{i+1}^2 \leqslant x_{i+2}x_{i-1}$$

or

$$x_i^2 \leqslant x_{i+2}x_{i-1} \qquad \text{and} \qquad x_{i+1}^2 \geqslant x_{i+2}x_{i-1}\,.$$

This may be expressed more succinctly by saying that $x_{i+2}x_{i-1}$ lies between the smaller and the larger of the numbers x_i^2 and x_{i+1}^2; or, since the x's are positive, that the geometric mean $\sqrt{x_{i-1}x_{i+2}}$ of x_{i-1} and x_{i+2} lies between x_i and x_{i+1}:

$$(2) \quad \min(x_i, x_{i+1}) \leqslant \sqrt{x_{i-1}x_{i+2}} \leqslant \max(x_i, x_{i+1})\,, \quad i = 1, 2, \dots, 5\,.$$

We shall show that the only solutions of this system of inequalities are $x_1 = x_2 = x_3 = x_4 = x_5$. The proof depends on the way the inequalities link the numbers x_i together. If we think of the x_i as sitting next to each other in a circle so that their indices increase as we go around the circle, then x_i and x_{i+1} are neighbors, and x_{i-1} and x_{i+2} are the "other" neighbors of x_i, x_{i+1}, respectively. The inequalities tell us that the geometric mean of x_{i-1}, x_{i+2} lies between the values of the neighbors x_i and x_{i+1}.

We may assume that

$$x_1 = \max\{x_1, x_2, x_3, x_4, x_5\} .$$

We consider the following two cases: (i) The second largest x_i is not a neighbor of x_1; (ii) The second largest x_i is a neighbor of x_1.

Case (i). Suppose for definiteness that max $\{x_2, x_3, x_4, x_5\} = x_4$. Then inequality (2), with $i = 2$, states that

$$\sqrt{x_1 x_4} \leqslant \max(x_2, x_3) . \tag{3}$$

Since x_1 and x_4 are the two largest among the x_i, (3) can hold only if $x_1 = x_4$ and either x_2 or x_3 is also equal to x_1. In any case, three of our five numbers are maximal (say x_1, x_4, x_2), so at most two (say x_3, x_5) are not maximal; those two are not neighbors. If $x_2 = x_1$ we have, by (2) with $i = 1$, $\min(x_1, x_2) = x_1 \leqslant \sqrt{x_5 x_3}$, and this holds only if $x_3 = x_5 = x_1$. Similarly if x_3 is maximal, i.e. $x_3 = x_4$, we have, by (2) with $i = 3$ $\min(x_3 x_4) = x_4 \leqslant \sqrt{x_2 x_5}$ which holds only if $x_2 = x_5 = x_4$. So all five of our numbers are equal.

Case (ii). Suppose for definiteness that max $\{x_2, x_3, x_4, x_5\} = x_5$. Inequality (2) with $i = 5$ states that the geometric mean of the non-neighbors x_2 and x_4 lies between the values of the neighbors x_1 and x_5, the two largest in our set:

$$x_5 = \min(x_5, x_1) \leqslant \sqrt{x_4 x_2} \ .$$

This implies that $x_2 = x_4 = x_5$. Thus $x_4 = \max\{x_2, x_3, x_4, x_5\}$, and we conclude as in Case (i) that $x_1 = x_2 = x_3 = x_4 = x_5$.

Note that our proof would work equally well if, instead of the geometric mean of non-neighbors x_{i-1}, x_{i+2}, any other means were used. The only property of a *mean* M of two numbers A, B needed in this proof is that it lies between the two numbers: $\min(A, B) \leqslant M \leqslant \max(A, B)$.

1972/5 First solution. Since $|f(x)|$ is bounded, it has a least upper bound M, and since f is not identically zero, $M > 0$.

Now suppose that the inequality $|g(y)| \leqslant 1$ does not hold for all y. Then there is a point y_0 such that $|g(y_0)| > 1$. Using the given equation and the triangle inequality, we get

$$2|f(x)||g(y_0)| = |f(x+y_0) + f(x-y_0)|$$
$$\leqslant |f(x+y_0)| + |f(x-y_0)| \leqslant 2M .$$

Hence

$$|f(x)| \leqslant \frac{M}{|g(y_0)|} < M ,$$

contradicting the fact that M is the least upper bound of $|f(x)|$. We conclude that $|g(y)| \leqslant 1$ for all y.

Second solution (due to Gerhard Arenstorf). Since f does not vanish identically, there is a number a such that $f(a) \neq 0$. We set $x = a + ny$ in the given equation and obtain, after transposing the term on the right side,

$$(1) \qquad f(a+(n+1)y) - 2f(a+ny)g(y) + f(a+(n-1)y) = 0 .$$

Consider y fixed, and abbreviate $f(a+ny)$ by f_n. Then (1) becomes the difference equation

$$(2) \qquad f_{n+1} - 2gf_n + f_{n-1} = 0$$

with solutions of the form

$$(3) \qquad f_n = b_1 r_1^n + b_2 r_2^n ,$$

provided the roots r_1, r_2 of the related quadratic equation†

$$r^2 - 2gr + 1 = 0$$

are distinct. Now $r_1 = g + \sqrt{g^2-1}$, $r_2 = g - \sqrt{g^2-1}$, and these are distinct if $g \neq \pm 1$.

If at some point y_0, $g(y_0) > 1$, then the above analysis for $y = y_0$ shows that $r_1 > 1$, and if $b_1 \neq 0$, f_n becomes unbounded as $n \to \infty$, contrary to hypothesis. If $b_1 = 0$, then $b_2 \neq 0$ (since $f(a) = b_2 \neq 0$.). Since $r_1 r_2 = 1$, we have $0 < r_2 < 1$. In this case (3) tells us that f_n is unbounded as $n \to -\infty$, contradicting the fact that $|f(x)| \leqslant 1$ for all x. A similar contradiction results if $g(y_0) < -1$ for some point y_0. We conclude that $|g(y)| \leqslant 1$.

Remark 1. We observe that both solutions remain valid if the given condition $|f(x)| \leqslant 1$ is replaced by $|f(x)| \leqslant B$, B any positive constant. Only the boundedness of f was used.

†See the second solution of 1963/4, specially eq. (16). Here the analogous characteristic equation, derived from recursion (2) with associated matrix $M = \begin{pmatrix} 0 & 1 \\ -1 & 2g \end{pmatrix}$ is $r^2 - 2gr + 1 = 0$; r_1 and r_2 are the eigenvalues of M. Or see S. Goldberg's *Introduction to Difference Equations*, John Wiley and Sons, Inc. (1958).

Remark 2. Readers who have studied some calculus may note that if $g(y)$ is twice differentiable in an interval including the origin, then it can be shown that f is twice differentiable everywhere, and that it satisfies the differential equation

$$f''(x) = g''(0)f(x) . \tag{4}$$

If $g''(0) \geqslant 0$, no non-trivial solution of (4) is bounded, so $g''(0) < 0$, in which case the general solution of (4) is

$$f(x) = A \sin(\alpha x + \theta) , \qquad \text{where } \alpha = \sqrt{-g''(0)} .$$

From this, we can deduce that $g(y) = \cos \alpha y$. Indeed the given equation is then satisfied; for

$A \sin[\alpha(x + y) + \theta] + A \sin[\alpha(x - y) + \theta] = 2A \sin(\alpha x + \theta) \cos \alpha y$, as the reader can verify by using the addition formulas.

Remark 3. Observe that the recursion (2) is satisfied by the functions $f_j = \cos jx$, $g = \cos x$, as can be verified by expressing

$$f_{n+1} = \cos(nx + x) , \qquad f_{n-1} = \cos(nx - x)$$

by the addition formulas for the cosine and adding them. With $f_0 = 1, f_1 = g$, the recursion generates $f_n = -f_{n-2} + 2gf_{n-1} = \cos nx$ for $n = 2, 3, \ldots$ as polynomials in $g = \cos x$. These are called Chebyshev polynomials and have important uses.

1972/6. Denote the planes, in order, by $\Pi_0, \Pi_1, \Pi_2, \Pi_3$, and let d_i be the distance from Π_i to Π_0 $(i = 1, 2, 3)$. We shall solve the problem by taking a regular tetrahedron $P_0P_1P_2P_3$ of edge length 1 and passing parallel planes $\Pi_0', \Pi_1', \Pi_2', \Pi_3'$ through its vertices in such a way that P_1, P_2, P_3 all lie on the same side of Π_0', and the distances d_i' between Π_i' and Π_0' satisfy

$$\frac{d_1'}{d_1} = \frac{d_2'}{d_2} = \frac{d_3'}{d_3} . \tag{1}$$

Once this is achieved, the configuration can be rotated so that $\Pi_0' \| \Pi_0$ and then enlarged (or shrunk) uniformly by the factor d_1/d_1'. The resulting regular tetrahedron of edge length d_1/d_1' is the one whose existence we were asked to establish.

It is convenient to place P_0 on plane Π_0 and to take P_0 as origin of a coordinate system in 3-space. We denote the unit vectors to P_1, P_2, P_3 again by P_1, P_2, P_3. We shall determine the planes Π_i' by finding the direction of their unit normal N. Since N is a unit vector perpendicular to Π_i', and P_i lies on Π_i', the distance from Π_i' to Π_0' is $d_i' = P_i \cdot N$.†

†$A \cdot B$ denotes the dot product or scalar product of vectors A and B; See p. 198

Consequently equations (1) may be written in the form

$$\frac{P_1 \cdot N}{d_1} = \frac{P_2 \cdot N}{d_2} = \frac{P_3 \cdot N}{d_3}, \tag{2}$$

where the coordinates (a_i, b_i, c_i) of the unit vectors P_i are known, and where the coordinates (x, y, z) of the normal vector N to Π_i' are to be determined. Equations (2) are equivalent to a system of two homogeneous equations

$$\begin{aligned} \alpha_1 x + \beta_1 y + \gamma_1 z &= 0 \\ \alpha_2 x + \beta_2 y + \gamma_2 z &= 0 . \end{aligned} \tag{3}$$

Here

$$\alpha_1 = \frac{a_1}{d_1} - \frac{a_2}{d_2}, \quad \beta_1 = \frac{b_1}{d_1} - \frac{b_2}{d_2}, \quad \gamma_1 = \frac{c_1}{d_1} - \frac{c_2}{d_2},$$

$$\alpha_2 = \frac{a_1}{d_1} - \frac{a_3}{d_3}, \quad \beta_2 = \frac{b_1}{d_1} - \frac{b_3}{d_3}, \quad \gamma_2 = \frac{c_1}{d_1} - \frac{c_3}{d_3}.$$

Such a system (3) always has at least one non-trivial solution (x, y, z). This is an instance of the general theorem that a system of $n - 1$ linear homogeneous equations in n unknowns always has a nontrivial solution. It can be proved for example inductively on n, by elimination.

Fifteenth International Olympiad, 1973

1973/1. We shall prove this assertion by induction on n; the proof is valid even if the region where the points P_i are permitted to lie includes the line g. We shall need the

LEMMA. *If vectors v and w make an angle $\theta \leqslant \pi/2$ with each other, then $|v + w| \geqslant |v|$ and $|v + w| \geqslant |w|$.* See Fig. 42*a*.

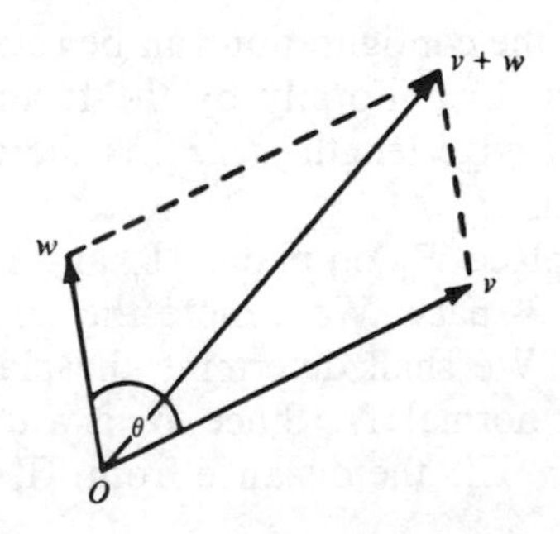

Figure 42a

We prove the lemma by writing the square of the length of a vector as the dot product of that vector with itself:

$$|v + w|^2 = (v + w) \cdot (v + w) = |v|^2 + |w|^2 + 2v \cdot w$$
$$= |v|^2 + |w|^2 + 2|v||w| \cos \theta \ .$$

Since the last term in this expression is non-negative for $0 \leqslant \theta \leqslant \pi/2$

$$|v + w|^2 \geqslant |v|^2 \qquad \text{and} \qquad |v + w|^2 \geqslant |w|^2 \ ,$$

and the lemma follows.

We shall use another immediate consequence of the parallelogram law of vector addition: that the vector $v + w$ lies between vectors v and w.

Clearly, we can name the points P_i so that they. are ordered in the counterclockwise direction with increasing index i along the semicircle with center O; see Fig. 42b. We denote OP_i by u_i and the vector sum $\Sigma_1^n u_i$ by u. We must show that for n odd, $|u| \geqslant 1$. (For an even number of u_i the theorem is obviously false; the unit vectors $u_1 = (12/13, 5/13)$, $u_2 = (-12/13, 5/13)$ both in the upper half plane, with $|u_1 + u_2| = |(0, 10/13)| < 1$ furnish a counterexample.)

The assertion holds for a single unit vector u_1, since $|u_1| = 1$. Suppose it holds for all sets containing an odd number less than n of unit vectors. We shall show that then it holds also for n (n odd) unit vectors.

Let $v = \Sigma_2^{n-1} u_i$ be the vector sum of the $n - 2$ unit vectors u_2, $u_3, \cdots, u_{n-1}$. By the induction hypothesis, $|v| \geqslant 1$. Also v lies between u_2 and u_{n-1}, hence certainly within the angle P_1OP_n.

Let $w = u_1 + u_n$, so that $v + w = u = \Sigma_1^n u_i$. Clearly $w = 0$ only if $u_n = -u_1$, which means u_1 and u_n lie along line g in opposite directions, and $|u| = |v + w| = |v| \geqslant 1$; so the assertion is true in that case. If $u_n \neq -u_1$, the vector w bisects $\angle P_1OP_n$ which is less than π, so w makes angles $< \pi/2$ with u_1 and u_n. Since v lies somewhere within $\angle P_1OP_n$, v and w make an acute angle with each other. Hence by the lemma,

$$|u| = |v + w| \geqslant |v| \geqslant 1 \ .$$

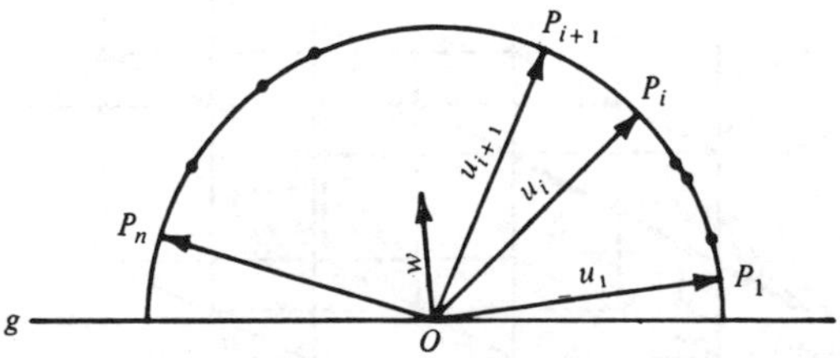

Figure 42b

1973/2. We shall answer this question by actually exhibiting sets of points with the desired properties. We describe below two sets, one consisting of 10 points, and one consisting of 27 points. We suggest that the ambitious reader try either to find such a set M with fewer than 10 points or prove that M must have at least 10 points.

Configuration M_1 consists of the eight corners of a cube and the two additional points obtained by reflecting the center of the cube in a pair of parallel faces; see Figure 43a. It is easy to verify that M_1 satisfies all the conditions of the problem.

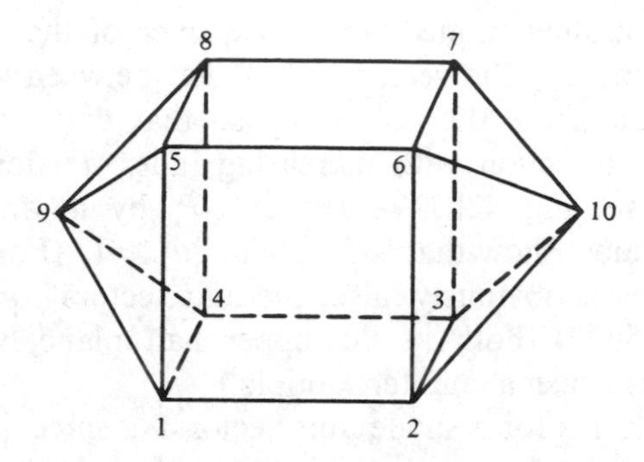

Figure 43a

Configuration M_2 consists of the center and the 26 lattice points on the boundary of the $2 \times 2 \times 2$ cube shown in Figure 43b. An equivalent description of M_2 is the set of 27 points with coordinates (x_1, x_2, x_3) where each x_i is $-1, 0$ or 1.

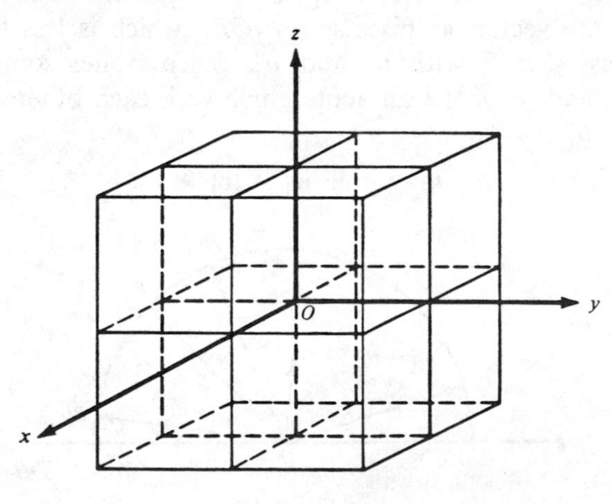

Figure 43b

Remark. If M is a point set satisfying the conditions of the problem, then any set affinely equivalent to M also satisfies them. For example, the set consisting of the vertices of two congruent hexagons lying in different planes which intersect in a i'long" diagonal of both hexagons is affinely equivalent to M_1 and thus also solves the problem, see Figure 43c.

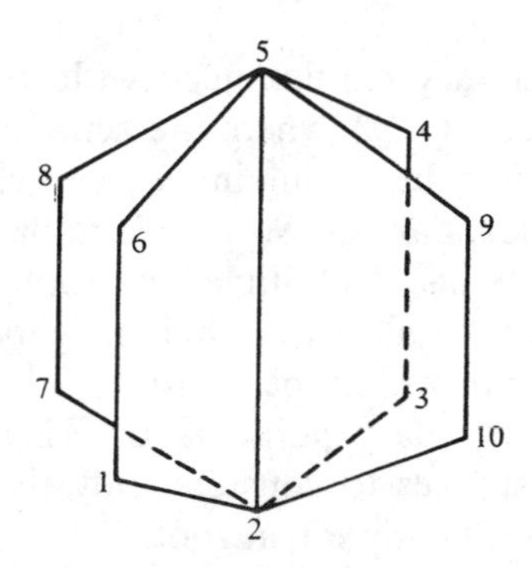

Figure 43c

1973/3 First solution (due to Paul Herdeg). First consider the equation

$$x + \frac{1}{x} = y\ , \qquad \text{for } y \text{ real}\ . \tag{1}$$

This is equivalent to $x^2 - yx + 1 = 0$, a quadratic in x which has real roots if and only if its discriminant $y^2 - 4$ is $\geqslant 0$, i.e. if and only if $|y| \geqslant 2$.

Now to solve $x^4 + ax^3 + bx^2 + ax + 1 = 0$, we divide it by x^2, then set $y = x + 1/x$ obtaining $y^2 + ay + (b - 2) = 0$. The roots of this quadratic equation are

$$y = x + \frac{1}{x} = \frac{-a \pm \sqrt{a^2 - 4(b-2)}}{2}\ ,$$

and the condition that at least one of them be $\geqslant 2$ in absolute value is then that

$$|a| + \sqrt{a^2 - 4(b-2)} \geqslant 4\ .$$

This is equivalent to $\sqrt{a^2 - 4(b-2)} \geqslant 4 - |a|$. After squaring both sides and subtracting a^2, we have $8|a| \geqslant 8 + 4b$; dividing this by 4 and squaring again yields $4a^2 \geqslant b^2 + 4b + 4$. Adding $4b^2$ to both sides, we have $4a^2 + 4b^2 \geqslant 5b^2 + 4b + 4$, so

$$a^2 + b^2 \geqslant \tfrac{5}{4}\left(b^2 + \tfrac{4}{5}b + \tfrac{4}{5}\right) = \tfrac{5}{4}\left(b + \tfrac{2}{5}\right)^2 + \tfrac{4}{5}\ .$$

The minimum value of the right member occurs when $b = -2/5$ and is $4/5$. Thus the minimum value of $a^2 + b^2$ is $4/5$.

Second solution. Another solution consists in employing a bit of useful point set topology to characterize the set S of admissible points (a, b), and then finding a point of S closest to the origin. To this end, we show that S is a closed set † by showing that its complement, the set T of points (a, b) for which $P(x) = 0$ has no real root, is open.† Indeed, if (a_0, b_0) belongs to T, all the zeros of $P(x)$ are complex, and since the zeros of a polynomial vary continuously with its coefficients, there is a whole neighborhood of (a_0, b_0) where the zeros of P remain complex.

Since S is closed, it indeed contains a point closest to the origin, and this point lies on the boundary of S. We claim that at any boundary point $(\bar{a}, \bar{b})$ of S, $P(x) = 0$ has a multiple real root. For if all its real roots there were simple, there would be a whole neighborhood of $(\bar{a}, \bar{b})$ where all real roots of $P = 0$ would be simple.‡ Thus $(\bar{a}, \bar{b})$ would be an interior point, not a boundary point of S. Thus, to find a point of S closest to $(0, 0)$, it suffices to consider only those values of (a, b) for which $P(x) = 0$ has a multiple real root.

We shall use the following property of the given polynomial P:

(2) If $P(r) = 0$, then $P\left(\frac{1}{r}\right) = 0$, since $P\left(\frac{1}{r}\right) = \frac{1}{r^4} P(r)$.

We distinguish two cases:

(a) The multiple root r of $P(x) = 0$ is neither 1 nor -1. Then, by property (2) above, $1/r \neq r$ is also a root; and since the product of all four roots is 1, the roots are

$$r, r, \frac{1}{r}, \frac{1}{r}.$$

Thus

$$\begin{aligned} P(x) &= (x-r)^2\left(x - \frac{1}{r}\right)^2 \\ &= x^4 - 2\left(r + \frac{1}{r}\right)x^3 + \left(r^2 + 2 + \frac{1}{r^2} + 2\right)x^2 - 2\left(r + \frac{1}{r}\right)x + 1 = 0. \end{aligned}$$

Setting $r + 1/r = y$, (see eq. (1)), we find that

$$a = -2y, \quad b = y^2 + 2, \tag{3}$$

where the real nature of r forces the condition $|y| \geqslant 2$. Hence $|a| \geqslant 4$,

†See NML vol. 18, *First Concepts of Topology*, by W. Chinn and N. Steenrod, p. 22.

‡ r is a simple root of $P(x) = 0$ if $x - r$ is a factor of $P(x)$ and $(x - r)^2$ is not. If the coefficients of P are varied sufficiently little, the roots vary so little that the root r retains its property of being a simple root.

$b \geqslant 6$. When y is eliminated from equations (3), we obtain the part

$$(4) \qquad b = \frac{a^2}{4} + 2, \qquad |a| \geqslant 4$$

of a parabola as boundary of S, see Figure 44.

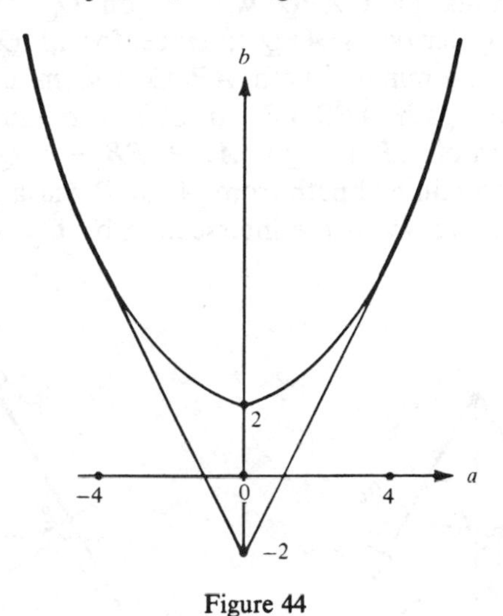

Figure 44

(b) The multiple root r of $P(x) = 0$ is 1 or -1. In this case, $P(1) = 2a + b + 2 = 0$ or $P(-1) = -2a + b + 2 = 0$ yield the symmetrically located pair of lines

$$(5) \qquad b = 2a - 2 \qquad \text{or} \qquad b = -2a - 2\ .$$

Every point (a, b) on these lines belongs to S, since $P(x) = 0$ has a real root 1 or -1 on them. Now the distance of lines (5) from $(0, 0)$ is

$$d = \frac{2}{\sqrt{5}} = \sqrt{a^2 + b^2}\ , \qquad \text{so that} \qquad a^2 + b^2 = \frac{4}{5}\ .$$

In case (a), on the other hand, we saw that $a^2 + b^2 \geqslant 4^2 + 6^2 = 52$. Therefore the minimum value of $a^2 + b^2$ is $4/5$.

1973/4. The region to be covered by the detector is pictured as $\triangle ABC$ in Figure 45a, the vertex A being the soldier's starting point, and h denoting the altitude of $\triangle ABC$. We are told that the detector's action ranges over a circle of radius $r = h/2$, but shall use this precise value of r only later. It is clear that in order to cover vertices B and C, the detector must, at some time, be placed on the circles O_B and O_C of radius r centered at B and C. We therefore shall first solve the simpler problem: What is the path from A of least length such that the detector checks points B and C? This is equivalent to the problem: What is the minimal path from A to a point Q on the circumference of O_B via a point P on the circumference of O_C? It will turn out that the solution of this problem is a solution of the given problem when $r = h/2$.

To find the minimal path APQ with P on O_C and Q on O_B, see Figure 45a, we consider paths $APQB$; since, for all Q on O_B, the piece QB has length r, the minimal path $APQB$ also minimizes the length of APQ. Conversely, a path APB (P on O_C) of minimal length furnishes the minimum length of $AP + PQ = AP + PB - r$, Q on O_B.

We claim that the minimal path from A to B via a point P on O_C is the path AP_0B, where P_0 is the intersection of O_C with the altitude of $\triangle ABC$ from C.

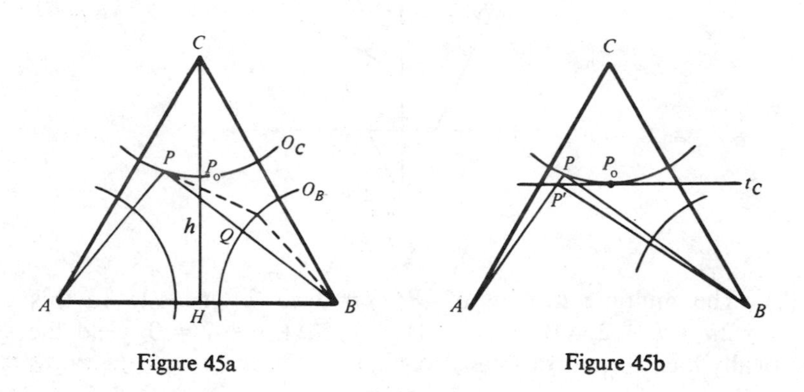

Figure 45a Figure 45b

PROOF: Denote by t_C the tangent to O_C at P_0, by P any other point on O_C, and by P' the intersection of AP with t_C; see Figure 45b. By the triangle inequality $P'B \leqslant P'P + PB$, so the length of $AP'B \leqslant$ the length of APB. This shows that the shortest path connecting A to B through a point P' on t_C is shorter than or equal to the shortest path connecting A to B through a point on P on O_C. Now the shortest path from A to B through a point on t_C has a well-known solution:† it is the path AP^*B, where P^* is the intersection of t_C with the perpendicular bisector of AB; see Fig. 45c.

In our configuration P^* is the point P_0 on the circumference of O_C. Thus the minimum problem involving t_C also solves the minimum problem involving O_C.

For $r = h/2$, the detector travelling along AP_0Q_0, where Q_0 is the intersection of P_0B with O_B, covers the entire region. We leave it to the

†This is known also as the reflection principle. See first solution of 1966/3, the reference given there, and Fig. 27c.

reader to check that no point of $\triangle ABC$ is farther than $h/2$ away from some point of the path.

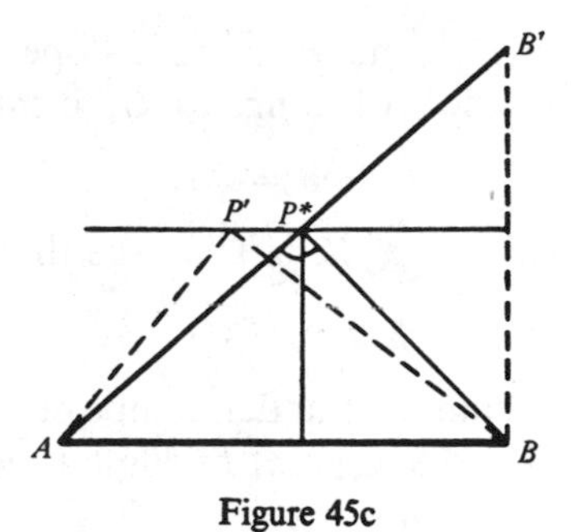

Figure 45c

It is easy to determine by the Pythagorean theorem that $AP_0^2 = 7h^2/12$, so that the minimal path has length

$$AP_0 + P_0Q_0 = AP_0 + P_0B - \frac{h}{2} = \left(\sqrt{\frac{7}{3}} - \frac{1}{2}\right)h \ ;$$

or, in terms of the side length s of $\triangle ABC$

$$AP_0 + P_0Q_0 = \left(\frac{\sqrt{7}}{2} - \frac{\sqrt{3}}{4}\right)s \ .$$

Remark. The method used in this solution serves also to solve the following, more general problem. Let K be a smooth closed convex curve, A and B points outside the region bounded by K. Find the shortest path APB, P on K. Here one must establish the existence of a point P_0 on K such that the angles made by AP_0 and P_0B with the normal n to the curve K are equal. The tangent to K at P_0 then plays the same role in the proof as the tangent t_C played in our solution.

1973/5. Parts (a) and (b) of the hypothesis may be summarized by saying that G is a group under thc operation of composition, the function $i(x) = x$ being the identity element, since $(f \circ i)(x) = (i \circ f)(x) = f(x)$. To each $f(x) = ax + b$ in G, we associate the coefficient a of x (the slope), and this association is multiplicative in the sense that for $f = ax + b$, $g = cx + d$, the coefficient associated with $g \circ f$ is ac.

Now every linear function $ax + b$ with $a \neq 1$ has a fixed point x_f; it is the solution $x_f = b/(1 - a)$ of the equation $ax_f + b = x_f$. The geometric characterization of x_f is the point where the line $y = ax + b$ crosses the line $y = x$. When $a = 1$, the function $ax + b = x + b$ has a fixed point if and only if $b = 0$; in this case the function is $i(x) = x$, and every point is a fixed point. Therefore hypothesis (c) bars all functions of slope 1 except the identity from the set G.

Now consider the composites $l = f\circ g$ and $m = g\circ f$ of the functions

(1). $$f = ax + b\ , \qquad g = cx + d \qquad \text{in } G\ .$$

Each has slope ac, while l^{-1} and m^{-1} have slope $1/ac$. It follows that $l^{-1}\circ m$ has slope 1, and since it belongs to G, it must be the identity:

$$l^{-1}\circ m = i\ .$$

Composing both sides with l on the left, we see that $m = l$, i.e.

(2) $$g\circ f = f\circ g\ .$$

Thus the functions in G commute under composition.

Now suppose that k is a fixed point of some function f in G. Then, for g in G,

(3) $$f[g(k)] = f\circ g(k) = g\circ f(k) = g[f(k)] = g(k)\ ,$$

so that $g(k)$ is also a fixed point of f. Either G contains only the identity function–in which case there is nothing to prove–or it contains a function $f(x) = ax + b$ with $a \neq 1$. Such a function f has the unique fixed point $k = b/(1 - a)$. Let g be any other function in G. Then by (3) $g(k) = k$ for every g in G, as was to be shown.

Alternatively, after showing that functions in G commute, we can substitute f and g as defined by (1) into (2) and find that $b/(1 - a) = d/(1 - c) = k$ is the unique fixed point of both.

Remark. It is easy to see that the set L of all non-constant linear functions forms a group under composition, and that this group L is *not* commutative. The solution of our problem shows that subgroups of L consisting of elements with fixed points are commutative, and that each such subgroup has, in fact, a common fixed point.

Conversely, consider the set of all linear non-constant functions $g_k(x)$ that leave a given number k fixed: $g_k(k) = k$. They form a group G_k since the composites and inverses of such functions also leave k fixed.

Moreover G_k is a commutative or *abelian* subgroup of L since, as we have seen, the fact that such groups have fixed points forces them to be commutative.

Let k and l be any two numbers, and let f be a linear function whose value at k is $l : f(k) = l$. It is easy to show–and we leave it to the reader–that if g_l is an element of G_l, then g_k defined by $g_k = f^{-1}\circ g_l\circ f$ is an element of G_k, and conversely. Symbolically we write, $G_k = f^{-1}\circ G_l\circ f$; subgroups related in this fashion are called *conjugate*.

1973/6. First we write condition (b) In the equivalent form

(b*) $$qb_k < b_{k+1} < \frac{1}{q}\, b_k\ .$$

It is simpler to solve the following slightly modified version of the problem: Let $a_1, a_2, \ldots, a_n$ be n non-negative numbers, and replace $<$ by $\leqslant$

in (a), (b*) and (c). The solution of this modified problem also solves the original problem, as we shall see.

We begin with the observation that if $(b_1, b_2, \ldots, b_n)$ solves our problem for a given vector $(a_1, a_2, \ldots, a_n)$, and $(b'_1, b'_2, \ldots, b'_n)$ solves it for another vector $(a'_1, a'_2, \ldots, a'_n)$, then $(b_1 + b'_1, b_2 + b'_2, \ldots, b_n + b'_n)$ solves the problem for the vector $(a_1 + a'_1, a_2 + a'_2, \ldots, a_n + a'_n)$. To verify this, we merely add the corresponding inequalities (a), (b*), (c) for the unprimed and primed quantities.

Next we observe that, if $(b_1, b_2, \ldots, b_n)$ solves our problem for a given vector $(a_1, a_2, \ldots, a_n)$, and if c is any positive constant, then $(cb_1, cb_2, \ldots, cb_n)$ solves the problem for the vector $(ca_1, ca_2, \ldots, ca_n)$. To verify this we merely multiply the inequalities (a), (b*), (c) by c.

Combining these observations we deduce that if we can solve the problem for several vectors $(a_1, a_2, \ldots, a_n)$, then we can solve it for any set obtained from them by addition and multiplication by positive constants. In particular, it suffices to solve the problem for the n unit vectors

$$u_1 = (1, 0, 0, \ldots 0)$$
$$u_2 = (0, 1, 0, \ldots 0)$$
$$\cdots\cdots\cdots\cdots\cdots\cdots$$
$$u_n = (0, 0, \ldots 0, 1)$$

because any other vector $(a_1, a_2, \ldots, a_n)$ with non-negative components can be obtained from these by addition and multiplication by positive constants.

The problem is easily solved for each of the n unit vectors above. For u_1, inequality (a) with $k = 1$ gives $b_1 = 1$ as the smallest possible value of b_1; all the other inequalities $a_k \leqslant b_k$, $k > 1$ are automatically satisfied for positive b_j. The smallest values for the b_j compatible with inequalities $qb_k \leqslant b_{k+1}$, see (b*), are

$$b_1 = 1\,, \qquad b_2 = q\,, \qquad b_3 = q^2\,, \qquad \ldots\,, \qquad b_n = q^{n-1}\,.$$

With this choice for the b's, inequality (c) is clearly satisfied since

$$\begin{aligned} b_1 + b_2 + \cdots + b_n &= 1 + q + q^2 + \cdots + q^{n-1} \\ &< 1 + q + q^2 + \cdots \\ &= \frac{1}{1-q} < \frac{1+q}{1-q} = \frac{1+q}{1-q}(1+0+0+\cdots+0)\,. \end{aligned}$$

Similarly, for the i-th unit vector $u_i = (0, 0, \ldots, 1, 0, \ldots)$ where all but the i-th component is 0 and $a_i = 1$, inequality (a), for $k = i$, gives $b_i = 1$ as the smallest possible value of b_i. The smallest values of b_k compatible with the inequalities (b*) are

$$b_k = q^{k-i} \quad \text{for } i < k\,, \qquad b_k = q^{i-k} \quad \text{for } k < i.$$

Now any admissible vector $(a_1, a_2, \ldots, a_n)$ can be written as

$$a_1u_1 + a_2u_2 + \cdots + a_nu_n,$$

so the corresponding solution vector is the sum

$$a_1(1, q, q^2, \ldots, q^{n-1}) + a_2(q, 1, q, \ldots, q^{n-2}) + \cdots + a_n(q^{n-1}, q^{n-2}, \ldots, q, 1) .$$

This leads to the following solution for any positive numbers $(a_1, a_2, \ldots, a_n)$:

$$(1) \quad \begin{aligned} b_1 &= a_1 + qa_2 + q^2a_3 + \cdots + q^{n-1}a_n \\ b_2 &= qa_1 + a_2 + qa_3 + q^2a_4 + \cdots + q^{n-2}a_n \\ b_3 &= q^2a_1 + qa_2 + a_3 + qa_4 + \cdots + q^{n-3}a_n \\ &\cdots\cdots\cdots\cdots\cdots\cdots \\ b_n &= q^{n-1}a_1 + q^{n-2}a_2 + \cdots + a_n . \end{aligned}$$

We shall now verify that the vector $(b_1, \ldots, b_n)$ defined by (1) is a solution of the original problem with the strict inequalities.

(a) Since all a_j are positive and q is positive, we have

$$(2) \quad b_k = q^{k-1}a_1 + \cdots + qa_{k-1} + a_k + qa_{k+1} + \cdots + q^{n-k}a_n > a_k$$

for $k = 1, 2, \ldots, n$; this proves (a).

(b) In addition to (2), the kth equation in system (1), we also write the $(k + 1) -$ st equation:

$$(3) \quad b_{k+1} = q^ka_1 + q^{k-1}a_2 + \cdots + qa_k + a_{k+1} + \cdots + q^{n-k-1}a_n .$$

Multiplying (2) by q yields

$$qb_k = q^ka_1 + \cdots + q^2a_{k-1} + qa_k + q^2a_{k+1} + \cdots + q^{n-k+1}a_n .$$

The first k terms of qb_k and of b_{k+1} given by (3) are equal, while since $q < 1$, the remaining terms of b_{k+1} are greater than those of qb_k. Hence $b_{k+1} > qb_k$, or equivalently

$$\frac{b_{k+1}}{b_k} > q .$$

Multiplying (2) by $1/q$ yields

$$\frac{1}{q}b_k = q^{k-2}a_1 + \cdots + a_{k-1} + \frac{1}{q}a_k + a_{k+1} + \cdots + q^{n-k-1}a_n .$$

Here the last $n - k$ terms are equal to those in (3), but the first k terms are greater than those in (3), so $b_k/q > b_{k+1}$, whence

$$\frac{b_{k+1}}{b_k} < \frac{1}{q} .$$

Thus we have shown that

$$q < \frac{b_{k+1}}{b_k} < \frac{1}{q} , \qquad k = 1, 2, \ldots, n-1 .$$

(c) By adding the n equations (1) we obtain a sum whose left side is $b_1 + b_2 + \cdots + b_n$, and whose right side is of the form

$$c_1a_1 + c_2a_2 + \cdots + c_na_n ,$$

where c_j is the sum of the coefficients of a_j in the j-th column of the right side of system (1). Now the sum of the n coefficients in any column is less than the number

$$c = 1 + 2q + 2q^2 + \cdots + 2q^{n-1} \tag{4}$$

obtained by including some additional powers of q. Thus

$$\begin{aligned} c_1a_1 + c_2a_2 + \cdots + c_na_n &< ca_1 + ca_2 + \cdots + ca_n \\ &= c(a_1 + a_2 + \cdots + a_n) . \end{aligned}$$

Moreover,

$$c = 1 + 2q(1 + q + q^2 + \cdots + q^{n-2}) \tag{5}$$

is less than the number we would get if we replaced the finite geometric series in the brackets of (5) with the infinite geometric series

$$1 + q + q^2 + \cdots = \frac{1}{1-q} ;$$

so

$$c < 1 + 2q\frac{1}{1-q} = \frac{1+q}{1-q} .$$

We therefore conclude that

$$b_1 + b_2 + \cdots + b_n < \frac{1+q}{1-q}(a_1 + a_2 + \cdots + a_n) .$$

Sixteenth International Olympiad 1974

1974/1. Let $N \geqslant 2$ be the number of rounds played. Then $N(p+q+r) = 39$. Since $0 < p < q < r$, we conclude that $p + q + r \geqslant 6$; and since 39 has prime factorization $39 = 3 \times 13$, it follows that

$$N = 3 , \qquad p + q + r = 13 .$$

Denoting by a_i, b_i, c_i the number of counters received by A, B, C in the i-th round, we tabulate the information we have so far, keeping in mind that each entry is one of the numbers p, q, r, and that the entries in each row are distinct:

game	A	B	C	total
1	a_1	b_1	c_1	13
2	a_2	b_2	c_2	13
3	a_3	r	c_3	13
total	20	10	9	39

Since $a_3 \neq r$, the sum of the numbers in the first column is at most $2r + q$. Hence

$$2r + q \geqslant 20 .$$

Substituting $q = 13 - p - r$ in this inequality, we obtain

$$r - p \geqslant 7 . \tag{1}$$

Since $p \geqslant 1$, this implies that $r \geqslant 8$.

On the other hand, the sum of the numbers in the second column is at least $r + 2p$. Hence

$$r + 2p \leqslant 10 , \tag{2}$$

which together with $p \geqslant 1$ implies that $r \leqslant 8$. Thus $r = 8$, and we can then conclude from either (1) or (2) that $p = 1$. Finally, $q = 13 - p - r = 4$. Since equality holds in (1) and (2), our bounds for the column sums are sharp. Hence the solution is

game	A	B	C
1	8	1	4
2	8	1	4
3	4	8	1

In particular, C receives $q = 4$ counters on the first round.

1974/2. Draw l through C parallel to AB; draw l' parallel to AB and so that AB lies midway between l and l'. Draw the circumcircle K of $\triangle ABC$; see Figure 46. We claim that the desired point D exists if and only if l' intersects (or touches) K. For, if Y is a point common to l' and K, then chord CY intersects chord AB in a point D, and $CD \cdot DY = AD \cdot DB$. Since C and Y are on l and l' while D is on AB, we have $DY = CD$, and so $CD^2 = AD \cdot DB$. If l' does not intersect K, there is no such point Y.

Denote by M the midpoint of the arc AB not containing C. Clearly l' intersects or touches K if and only if the distance from M to $AB \geqslant$ distance from C to AB, i.e. area $\triangle ABM \geqslant$ area $\triangle ABC$, or

$$\tfrac{1}{2} AM \cdot BM \sin(\pi - C) \geqslant \tfrac{1}{2} ab \sin C , \tag{1}$$

where a and b denote the lengths of BC and AC. Since $AM = BM$ and $\sin(\pi - C) = \sin C$, (1) reduces to

$$AM^2 \geqslant ab \,. \tag{2}$$

Let r be the radius of K and observe that

$$AM = 2r \sin \frac{C}{2} \,, \qquad a = 2r \sin A \,, \qquad b = 2r \sin B \,.$$

Substituting this into (2) and dividing by $4r^2$, we obtain the desired condition

$$\sin^2 \frac{C}{2} \geqslant \sin A \sin B \,.$$

Note that equality holds if and only if l' is tangent to K. In this case there is only one point D on AB such that $AD \cdot DB = CD^2$.

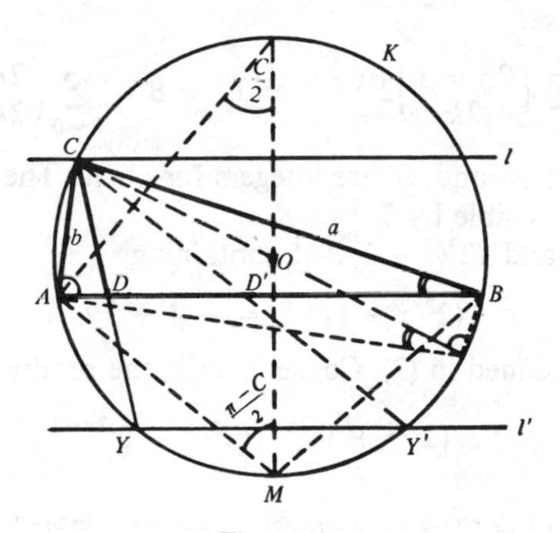

Figure 46

Remark. The necessity and sufficiency of the inequality $\sin A \sin B \leqslant \sin^2(C/2)$ also follows easily from a mild application of calculus to the function

$$f(\gamma_1) = \sin \gamma_1 \sin \gamma_2 = \sin \gamma_1 \sin(C - \gamma_1) \,,$$

which is differentiable in the interval $0 \leqslant \gamma_1 \leqslant C/2$. Its derivative is

$$\begin{aligned} f'(\gamma_1) &= \cos \gamma_1 \sin(C - \gamma_1) - \sin \gamma_1 \cos(C - \gamma_1) \\ &= \sin(C - \gamma_1 - \gamma_1) = \sin(C - 2\gamma_1) \,; \end{aligned}$$

$f'(\gamma_1)$ is positive for $0 \leqslant \gamma_1 < C/2$, and assumes all values between $f(0) = 0$ and $f(C/2) = \sin^2(C/2)$. In particular, there is a point γ_1 for which $f(\gamma_1) = \sin A \sin B$ if and only if $0 \leqslant \sin A \sin B \leqslant \sin^2(C/2)$.

1974/3 First solution. The binomial coefficients $\binom{2n+1}{2k+1}$ occur in the second, fourth, sixth, ... terms of the expansion of $(a + b)^{2n+1}$; and for

$a = 2^{3/2}$, $b = 1$, these alternate terms are

$$\binom{2n+1}{2k+1}2^{3(2k+1)/2} = \binom{2n+1}{2k+1}2^{3k}2^{3/2} \quad (k = 0, 1, \ldots, n) .$$

Accordingly, we expand $(2^{3/2} + 1)^{2n+1}$ and group together alternate terms, obtaining

$$(2^{3/2}+1)^{2n+1} = \binom{2n+1}{0} + \binom{2n+1}{2}2^3 + \binom{2n+1}{4}2^6 + \cdots + \binom{2n+1}{2n}2^{3n}$$

$$(1) \qquad + 2^{3/2}\left[\binom{2n+1}{1} + \binom{2n+1}{3}2^3\right.$$

$$\left. + \binom{2n+1}{5}2^6 + \cdots + \binom{2n+1}{2n+1}2^{3n}\right] = A + 2^{3/2}B ,$$

where we have set

$$(2) \qquad A = \sum_{k=0}^{n}\binom{2n+1}{2k}2^{3k} \quad \text{and} \quad B = \sum_{k=0}^{n}\binom{2n+1}{2k+1}2^{3k} .$$

We observe that A and B are integers for all n. The problem is to show that B is not divisible by 5.

Next, we expand $(2^{3/2} - 1)^{2n+1}$, obtaining

$$(2^{3/2} - 1)^{2n+1} = -A + 2^{3/2}B,$$

with A, B as defined in (2). Consequently, the product

$$(2^{3/2} + 1)^{2n+1}(2^{3/2} - 1)^{2n+1}$$

is

$$(3) \quad (2^3 - 1)^{2n+1} = 2^3B^2 - A^2 , \qquad \text{or} \qquad 7^{2n+1} = 8B^2 - A^2 .$$

The left side in the last expression,

$$7^{2n+1} = (7^2)^n \cdot 7 = (49)^n(5 + 2) = (50 - 1)^n(5 + 2) ,$$

has remainder $2(-1)^n$ on division by 5; i.e. $7^{2n+1} \equiv 2(-1)^n \pmod 5$. Therefore

$$(4) \qquad 8B^2 - A^2 \equiv 2(-1)^n \pmod 5 .$$

Now if B were divisible by 5, the congruence (4) would read

$$(4)' \qquad A^2 \equiv \pm 2 \pmod 5 .$$

We claim that the square of no integer is $\equiv \pm 2 \pmod 5$; for

$$0^2 \equiv 0, \quad 1^2 \equiv 1, \quad 2^2 \equiv -1, \quad 3^2 \equiv -1, \quad 4^2 \equiv 1 ,$$

modulo 5, and for any other $A = 5k + l$, we have

$$A^2 = (5k + l)^2 \equiv l^2 \pmod 5 , \quad \text{where } l = 0, 1, 2, 3, 4 .$$

Thus (4)′ is impossible, which shows that B is not divisible by 5.

Second solution. In the expansion (1), let us indicate the dependence of A and B on the integer n by attaching subscripts. For $n = 0$, $(2^{3/2}+1)^{2n+1} = (2^{3/2}+1)^1 = 1 + 2^{3/2}\cdot 1$; so

$$A_0 = 1, \qquad B_0 = 1 .$$

To derive recursion relations for A_j, B_j in the expression

$$(2^{3/2}+1)^{2j+1} = A_j + 2^{3/2}B_j ,$$

we write

$$\begin{aligned}(2^{3/2}+1)^{2j+1} &= (2^{3/2}+1)^2(2^{3/2}+1)^{2j-1}\\ &= (9+2\cdot 2^{3/2})\big(A_{j-1}+2^{3/2}B_{j-1}\big)\\ &= 9A_{j-1}+16B_{j-1}+(2A_{j-1}+9B_{j-1})2^{3/2} ;\end{aligned}$$

thus

$$(5) \qquad A_j = 9A_{j-1}+16B_{j-1} , \qquad B_j = 2A_{j-1}+9B_{j-1} .$$

We must prove that B_j is not divisible by 5 for any j. We reduce (5) modulo 5:

$$(5)' \qquad A_j \equiv -A_{j-1}+B_{j-1} , \qquad B_j \equiv 2A_{j-1}-B_{j-1} ,$$

and we must show that

$$B_j \equiv 2A_{j-1}-B_{j-1} \not\equiv 0 \pmod 5 .$$

At first glance, this looks like an infinite task; but since there are only 5 residue classes modulo 5, there are at most $5^2 = 25$ distinct pairs of numbers (A_j, B_j) modulo 5. So among the first 25 pairs obtained by the recursion formulas (5)′, a pair must be repeated and from then on, all subsequent pairs are repetitions of the previously generated pairs. It turns out that just three applications of the recursion relations suffice; for

$$\begin{cases} A_1 \equiv -A_0+B_0 \equiv -1+1 \equiv 0\\ B_1 \equiv \ \ 2A_0-B_0 \equiv \ \ 2-1 \equiv 1 ;\end{cases}$$

$$\begin{cases} A_2 \equiv -A_1+B_1 \equiv -0+1 \equiv 1\\ B_2 \equiv \ \ 2A_1-B_1 \equiv \ \ 0-1 \equiv -1 ;\end{cases}$$

$$\begin{cases} A_3 \equiv -A_2+B_2 \equiv -1-1 \equiv -2 \equiv 3\\ B_3 \equiv \ \ 2A_2-B_2 \equiv \ \ 2+1 \equiv 3 ;\end{cases}$$

so that $(A_3, B_3) \equiv 3(A_0, B_0)$. It follows that $(A_j, B_j) \equiv (3A_{j-3}, 3B_{j-3})$ for all $j \geqslant 3$, and since B_0, B_1, B_2 are not divisible by 5, neither are $B_3 \equiv 3B_0$, $B_4 \equiv 3B_1$, $B_5 \equiv 3B_2$, etc.

Remark. Recursion relations (5′) may be written in vector notation as

$$u_j \equiv Tu_{j-1}, \quad \text{where} \quad u_k = \begin{pmatrix} A_k \\ B_k \end{pmatrix} \quad \text{and } T = \begin{pmatrix} -1 & 1 \\ 2 & -1 \end{pmatrix};$$

see the second solution of 1963/4 for such a treatment of linear recursions. We can find u_j for $j = 1, 2, \ldots$ by using iterates of the matrix T:

$$u_j \equiv Tu_{j-1} \equiv T[Tu_{j-2}] \equiv T^2u_{j-2} \equiv \cdots \equiv T^ju_0 .$$

The reader may verify that $T^3 \equiv 3I$, where I is the identity matrix $\binom{10}{01}$. It follows that $A_3 \equiv 3A_0$, $B_3 \equiv 3B_0$

1974/4. Since the board has 32 white squares, we have

$$a_1 + a_2 + \cdots + a_p = 32 .$$

Since $a_1 \geqslant 1$, $a_2 \geqslant 2$, $\ldots$, $a_p \geqslant p$, we have

$$32 \geqslant 1 + 2 + \cdots + p = \tfrac{1}{2}p(p+1) .$$

It follows that $p^2 + p \leqslant 64$, so $p \leqslant 7$. This shows that there can be at most 7 rectangles in a decomposition of the type required.

To show that decompositions into 7 rectangles exist and to find them all, we seek seven distinct integers whose sum is 32. Here is a complete list:

(a) $1 + 2 + 3 + 4 + 5 + 6 + 11$

(b) $1 + 2 + 3 + 4 + 5 + 7 + 10$

(c) $1 + 2 + 3 + 4 + 5 + 8 + 9$

(d) $1 + 2 + 3 + 4 + 6 + 7 + 9$

(e) $1 + 2 + 3 + 5 + 6 + 7 + 8$

Now (a) is impossible because no rectangle with 11 white and 11 black squares, hence with $2 \times 11 = 22$ squares, can be cut from an 8×8 board. All other cases are possible and corresponding subdivisions are shown in Figures 47 b-e. Thus cases (b)-(e) solve the problem.

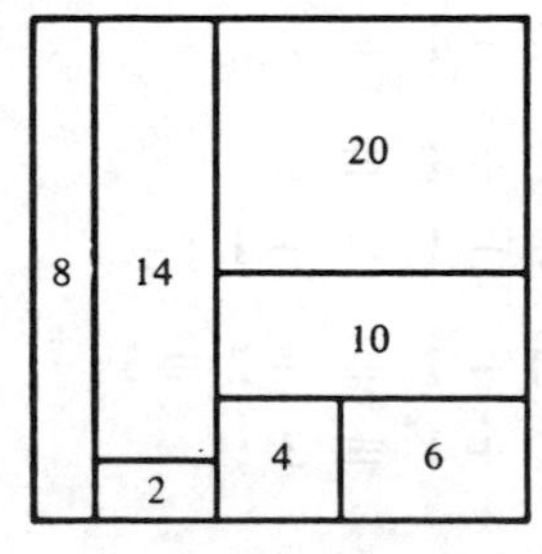

Figure 47b

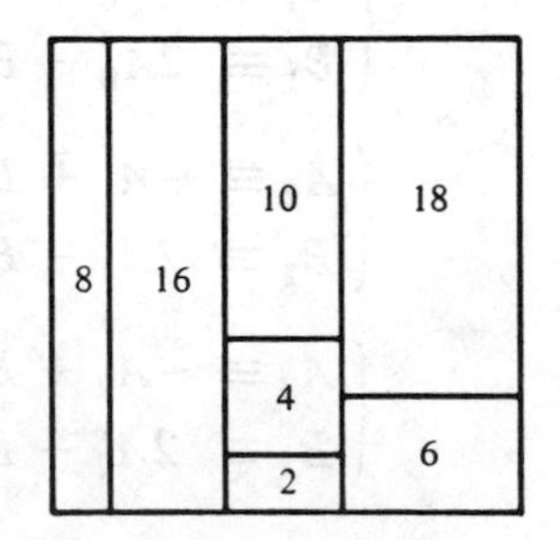

Figure 47c

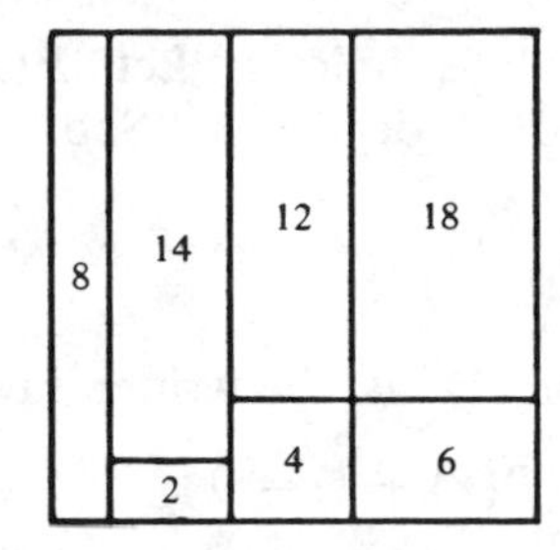

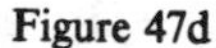
Figure 47d

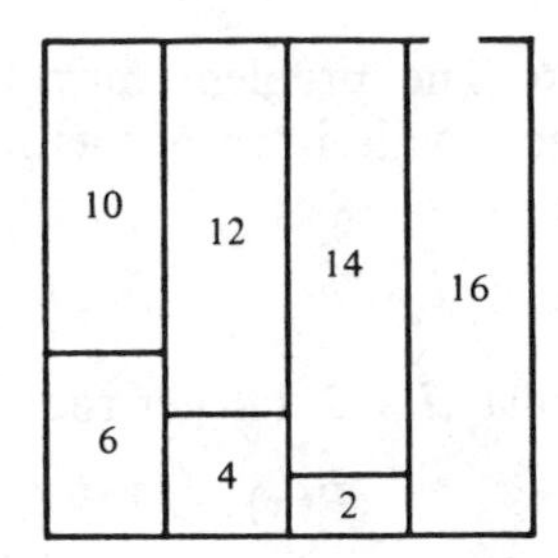

Figure 47e

1974/5. Clearly

$$S > \frac{a}{a+b+c+d} + \frac{b}{a+b+c+d} + \frac{c}{a+b+c+d} + \frac{d}{a+b+c+d} = 1 .$$

Suppose for definiteness that d is the largest of the four numbers a, b, c, d. Then

$$S \leqslant \frac{a}{a+b+c} + \frac{b}{a+b+c} + \frac{c}{a+b+c} + \frac{d}{a+c+d}$$

$$= 1 + \frac{d}{a+c+d} < 1 + 1 = 2 .$$

Thus $1 < S < 2$.

We shall show next that S actually assumes all values in the open interval $(1, 2)$. First we observe that S varies continuously with the positive numbers a, b, c, d. Hence if we can show that S takes values arbitrarily close to the endpoints 1 and 2 of this interval, it will follow that S assumes all values between 1 and 2.

Set $a = 1$, $b = \varepsilon$, $c = d = \varepsilon^2$, where $\varepsilon > 0$. Then

$$S = \frac{1}{1+\varepsilon+\varepsilon^2} \frac{\varepsilon}{1+\varepsilon+\varepsilon^2} + \frac{\varepsilon^2}{\varepsilon+2\varepsilon^2} + \frac{\varepsilon^2}{1+2\varepsilon^2} .$$

As $\varepsilon \to 0$, the first term approaches 1, the others approach 0, so $S \to 1$.

Next, set $a = c = 1$, $b = d = \varepsilon$; then

$$S = \frac{1}{1+2\varepsilon} + \frac{\varepsilon}{2+\varepsilon} + \frac{1}{1+2\varepsilon} + \frac{\varepsilon}{2+\varepsilon} .$$

As $\varepsilon \to 0$, the first and third fractions approach 1, while the second and fourth approach 0; hence $S \to 2$.

Remark. Note that the cyclic permutation $abcd \to bcda$ takes each term of S into the next term; for this reason S is called a cyclic sum. The method used to solve the above problem can be extended to find the ranges of more general cyclic sums.

1974/6. The problem may be restated as follows: Let $P(x)$ be a polynomial with integer coefficients and of degree d. Show that the equation

(1) $$[P(x)]^2 = 1$$

has at most $d + 2$ integer roots. Equation (1) may be written in the form

(1′) $$P(x)^2 - 1 = (P(x) - 1)(P(x) + 1) = 0\ .$$

Set $Q(x) = P(x) - 1$; then $Q(x) + 2 = P(x) + 1$, and the problem becomes: Let $Q(x)$ be a polynomial of degree d with integer coefficients. Show that the equation

(2) $$Q(x)(Q(x) + 2) = 0$$

has at most $d + 2$ integer roots; that is, show that $Q(x)$ and $Q(x) + 2$ have a total of at most $d + 2$ integer zeros. We shall use the following

LEMMA. *Let m be an integer zero of the polynomial $F(x)$ with integer coefficients. Then the only possible integer zeros of the polynomials $F(x) + p$ or $F(x) - p$, where p is a prime, are $m - p$, $m - 1$, $m + 1$, $m + p$.*

PROOF: Since m is a zero of $F(x)$, $x - m$ is a factor. Thus $F(x) = (x - m)G(x)$ and so

$$F(x) \pm p = (x - m)G(x) \pm p\ ,$$

where G is a polynomial with integer coefficients. Hence if $F(x) \pm p = 0$ for some integer x, then $(x - m)G(x) \pm p = 0$, or

$$(x - m)G(x) = \mp p\ .$$

Thus $x - m$ divides p, and since the only divisiors of p are ± 1 and $\pm p$, we have $x - m = \pm 1$ or $\pm p$. Hence $x = m + 1$ or $m - 1$ or $m + p$ or $m - p$. This completes the proof of the lemma.

We now apply this lemma (with $p = 2$) to the situation where m is the *least* integral zero of $Q(x)(Q(x) + 2)$, assuming that it has an integral zero. Say for definiteness that m is a zero of $Q(x)$. Then by the lemma, the only possible zeros of $Q(x) + 2$ are $m + 1$ and $m + 2$. Since $Q(x)$ has at most d zeros, the product $Q(x)(Q(x + 2)$ has at most $d + 2$ integral zeros.

Remark. Our proof actually shows that if both $Q(x)$ and $Q(x) + 2$ have integral zeros, then the total number of such zeros is at most 4. For if we normalize as above, then all integral zeros of $Q(x) + 2$ are $\leqslant m + 2$, and hence all integral zeros of $Q(x)$ are $\leqslant m + 4$. But if both m and $m + 4$ are roots of $Q(x) = 0$, then $Q(x) = (x - m)(x - m - 4)R(x)$. If x is an integer, then the first two of these three factors are 4 units apart, so their product cannot be -2. Thus there is no integer x such that $Q(x) = -2$. Hence all the zeros of $Q(x)(Q(x) + 2)$

are in the set $\{m, m+1, m+2, m+3\}$. With the help of this fact, it is a simple matter to show that the upper bound $d+2$ in the statement of the problem can be improved to the function $f(d)$ defined as follows:

d	1	2	3	$\geqslant 4$
$f(d)$	2	4	4	d .

The following examples show that this improved bound is actually sharp.

$$d = 1 \quad P(x) = x$$

$$d = 2 \quad P(x) = x^2 + x - 1$$

$$d = 3 \quad P(x) = x^3 - x^2 - 2x + 1$$

$$d \geqslant 4 \quad P(x) = (x-1)(x-2)\ldots.(x-d) + 1$$

Seventeenth International Olympiad, 1975

1975/1. Since

$$\sum_{i=1}^{n} (x_i - y_i)^2 = \sum_{i=1}^{n} x_i^2 - 2\sum_{i=1}^{n} x_i y_i + \sum_{i=1}^{n} y_i^2$$

and

$$\sum_{i=1}^{n} (x_i - z_i)^2 = \sum_{i=1}^{n} x_i^2 - 2\sum_{i=1}^{n} x_i z_i + \sum_{i=1}^{n} z_i^2 ,$$

and since $\Sigma_{i=1}^{n} y_i^2 = \Sigma_{i=1}^{n} z_i^2$, the inequality to be proved is equivalent to

$$(1) \qquad \sum_{i=1}^{n} x_i y_i \geqslant \sum_{i=1}^{n} x_i z_i .$$

In the sum on the left, the factors y_i occur in descending order, while in the sum on the right, the factors z_i may not occur in descending order. That is, there may be terms $x_k z_k$ and $x_l z_l$ with $k < l$ such that

$$(2) \qquad x_k \geqslant x_l \qquad \text{and} \qquad z_k < z_l .$$

Whenever this is the case, we interchange z_k and z_l, replacing the terms $x_k z_k + x_l z_l$ by $x_k z_l + x_l z_k$. We claim that such a replacement does not decrease the sum on the right of (1). For the product

$$(x_k - x_l)(z_k - z_l) = x_k z_k + x_l z_l - (x_k z_l + x_l z_k)$$

is $\leqslant 0$ in view of (2), so that

$$x_k z_k + x_l z_l \leqslant x_k z_l + x_l z_k .$$

After a finite number of such interchanges, the sum on the right of (1) is replaced by a sum $\Sigma_{i=1}^{n} x_i z_i'$ at least as large, where $z_1' \geqslant z_2' \geqslant \ldots \geqslant z_n'$. At this point we have $z_i' = y_i$. Thus the sum on the right side of (1) is $\leqslant \Sigma_{i=1}^{n} x_i y_i$, which proves the assertion.

1975/2. For $0 \leqslant r < a_p$, denote by B_r the subsequence of $a_1, a_2, a_3, \ldots$ consisting of those members which are congruent to r modulo a_p. Since there are only finitely many remainders upon division by a_p (namely $0, 1, \ldots, a_p - 1$) while the sequence $a_1, a_2, \ldots$ is infinite, there is at least one value of r for which B_r has infinitely many members. Let r be such a value.

Suppose a_q is the smallest member of B_r which exceeds a_p. Then for all a_m in B_r with $m > q$, we have

$$a_m - a_q \equiv 0 (\bmod a_p) , \qquad \text{i.e.} \quad a_m - a_q = xa_p ,$$

where x is an integer. Thus

$$a_m = xa_p + a_q .$$

This solves the problem with $x = (a_m - a_q)/a_p$ and $y = 1$.

1975/3 First solution. Figure 48a shows the construction described in the problem. We note that $\angle AQC = \angle BPC = 105°$, and $\angle ARB = 150°$. Construct $RX = RA$ and perpendicular to RB, and draw AX and QX. Since $\angle ARB = 150°$ and $\angle XRB = 90°$, $\angle ARX = 60°$. Thus isosceles triangle ARX $(RX = RA)$ has a vertex angle of $60°$ and hence is equilateral.

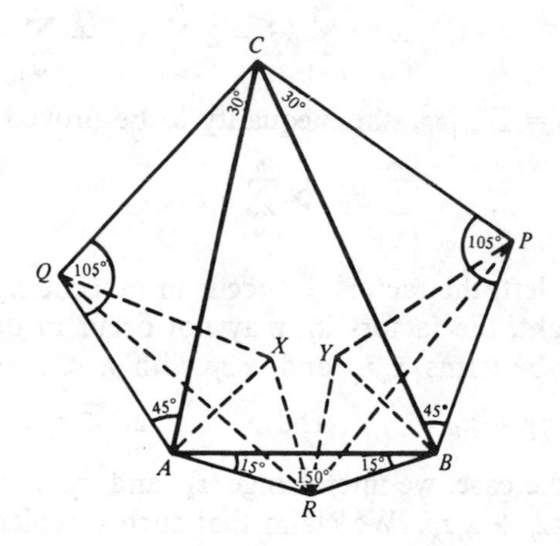

Figure 48a

Now $\angle BAQ = \angle A + 45°$ and $\angle BAX = 60° - 15° = 45°$, so $\angle XAQ = \angle BAQ - \angle BAX = \angle A$. In $\triangle ACQ$, by the law of sines,

$$\frac{AQ}{AC} = \frac{\sin 30°}{\sin 105°} = \frac{\sin 30°}{\sin 75°} = \frac{2 \sin 15° \cos 15°}{\cos 15°} = 2 \sin 15° ,$$

and in $\triangle ABR$,

$$\frac{AR}{AB} = \frac{\sin 15°}{\sin 150°} = \frac{\sin 15°}{\sin 30°} = 2 \sin 15° .$$

It follows that

$$(1) \qquad \frac{AR}{AB} = \frac{AX}{AB} = \frac{AQ}{AC} ,$$

so $AX/AQ = AB/AC$ and $\triangle AXQ \sim \triangle ABC$. Therefore $\angle AQX = \angle C$, $\angle AXQ = \angle B$.

Since $\angle RBP = 15° + \angle B + 45° = 60° + \angle B$ and $\angle RXQ = 60° + \angle B$, we have $\angle RBP = \angle RXQ$.

As before, construct $RY = RB$ and perpendicular to RA. Again, we have an equilateral $\triangle BRY$, and $\triangle YBP \sim \triangle ABC$, by the same reasoning as before. We deduce that $\triangle AXQ \sim \triangle YBP$; and since their corresponding sides AX and YB are equal, triangles AXQ and YBP are, in fact, congruent. Since the attached equilateral triangles ARX and YRB are also congruent, we see that

$$\text{quadrilateral } ARXQ \cong \text{quadrilateral } YRBP \ .$$

It follows that corresponding diagonals RQ and RP are equal. Moreover, since we constructed $XR \perp RB$, rotating $RAQX$ about R by 90° clockwise will bring it into coincidence with $RYPB$. Hence RQ and RP must be perpendicular.

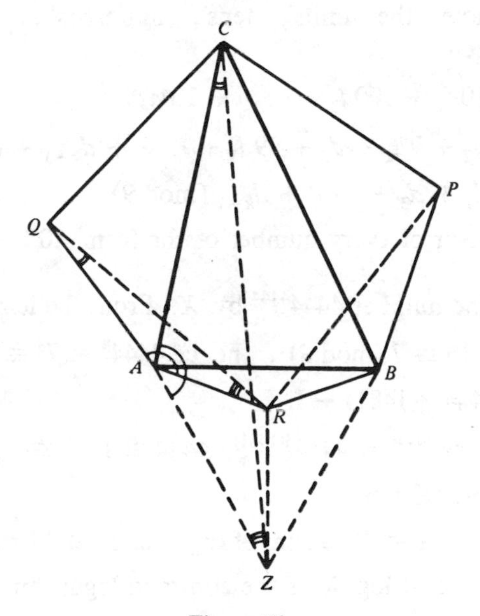

Figure 48b

Second solution. On side AB of $\triangle ABC$, construct equilateral $\triangle ABZ$ externally, see Figure 48b, and draw RZ and CZ. Now $\angle ZAR = 60° - 15° = 45°$, so $\angle QAR = \angle CAZ$; and as we showed in the first solution, see (1), $AQ/AR = AC/AB$. Since $AB = AZ$, it follows that

(2) $\triangle AQR \sim \triangle ACZ$. Similarly, $\triangle BPR \sim \triangle BCZ$.

Moreover, since triangles CAQ and CBP are similar,

$$\frac{AC}{AQ} = \frac{BC}{BP} = k ,$$

so the constant of magnification is the same for both pairs of triangles in (2).

Now rotate $\triangle AQR$ clockwise 45° about A and dilate in the ratio k; rotate $\triangle BPR$ counterclockwise 45° about B and dilate in the ratio k. These transformations take QR into CZ and PR into CZ, so the pre-images QR and PR are equal and make an angle of $2 \cdot 45° = 90°$ with each other.

1975/4. We shall solve this problem using some rudimentary facts about common logarithms and the following

LEMMA. *Every integer n, written in decimal notation, is congruent to the sum of its digits modulo 9.*†

PROOF: Denote the units', tens', hundreds'... digits of n by $d_1, d_2, \ldots$. Then

$$\begin{aligned} n &= d_1 + 10d_2 + 100d_3 + \cdots + 10^k d_{k+1} \\ &= d_1 + d_2 + 9d_2 + d_3 + 99d_3 + \cdots + d_{k+1} + (10^k - 1)d_{k+1} \\ &\equiv d_1 + d_2 + d_3 + \cdots + d_{k+1} \pmod 9 , \end{aligned}$$

since 9 is a divisor of every number of the form $10^m - 1$.

Abbreviate the number 4444^{4444} by X. From the lemma we deduce that

$$4444 \equiv 16 \equiv 7 \pmod 9 , \quad \text{hence } 4444^3 \equiv 7^3 \equiv 1 \pmod 9 ;$$

and since $4444 = 3(1481) + 1$,

$$X = 4444^{4444} = 4444^{3(1481)} \cdot 4444 \equiv 1 \cdot 7 \equiv 7 \pmod 9 .$$

Thus the lemma tells us

$$X \equiv A \equiv B \equiv \text{sum of digits of } B \equiv 7 \pmod 9 .$$

On the other hand, if $\log X$ is the common logarithm of X, then

$$\log X = 4444 \log 4444 < 4444 \log 10^4 = 4444 \cdot 4 ,$$

that is

$$\log X < 17776 .$$

†See p. 197.

Now, if the common logarithm of an integer is less than c, that integer has at most c digits. So X has at most 17776 digits, and even if all of them were 9, the sum of the digits of X would be at most $9 \times 17776 = 159984$. Therefore

$$A \leqslant 159984 .$$

Among all natural numbers less than or equal to 159984, the one whose sum of digits is a maximum is 99999. It follows that $B \leqslant 45$; and among all natural numbers less than or equal to 45, 39 has the largest sum of digits, namely 12. So the sum of the digits of B is at most 12. But the only natural number not exceeding 12 and congruent to 7 modulo 9 is 7. Thus 7 is the solution to the problem.

1975/5. We shall prove that there are, in fact, infinitely many points on the unit circle such that the distances between any two of them are rational.† We shall construct these points as vertices of right triangles ABC with diameter AB of the unit circle as common hypotenuse; see Fig. 49a.

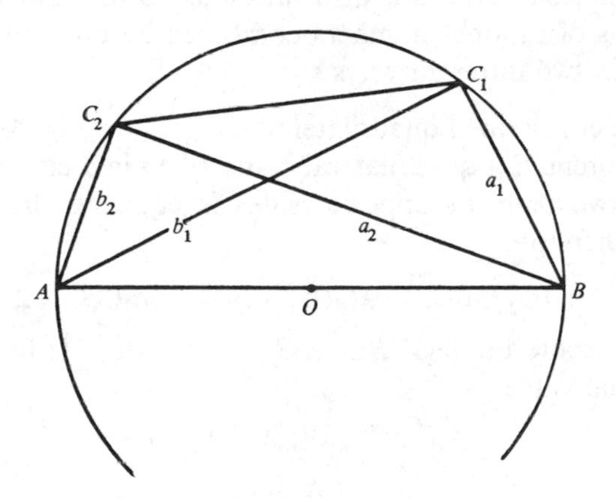

Figure 49a

Now a right triangle with rational sides a, b, c satisfies $a^2 + b^2 = c^2$ When we multiply a, b, c by their common denominator d, the resulting integers also satisfy the Pythagorean relation $(ad)^2 + (bd)^2 = (cd)^2$. A set of positive integers α, β, γ such that $\alpha^2 + \beta^2 = \gamma^2$ is called a *Pythagorean triple*. Infinitely many such triples can be generated by taking

†Thus this problem has nothing to do with the year 1975; it is a non-sectarian problem.

arbitrary natural numbers m and n and setting

$$(1) \qquad \alpha = 2mn\,, \quad \beta = m^2 - n^2\,;$$

then $$\gamma^2 = \alpha^2 + \beta^2 = (m^2 + n^2)^2\,, \quad \gamma = m^2 + n^2\,.$$

In our problem the hypotenuse is the diameter of the unit circle and therefore has length 2. We "normalize" α, β, γ by dividing each by $\frac{1}{2}(m^2 + n^2)$, obtaining the rational lengths

$$(2) \qquad a = \frac{4mn}{m^2 + n^2}\,, \qquad b = \frac{2(m^2 - n^2)}{m^2 + n^2}\,, \qquad c = 2\,.$$

Every distinct, relatively prime pair of natural numbers m, n thus yields a different triple $a, b, 2$; and since rational operations $(+, -, \times, \div)$ performed on rational numbers yield rational numbers, we can construct infinitely many right triangles whose legs have rational lengths a and b given by (2), and whose hypotenuses have length 2. Their vertices lie on the unit circle.

We show next that, if ABC_1 and ABC_2 are two right triangles with legs of rational lengths and hypotenuse AB of length 2, then the distance C_1C_2 between their vertices is also rational. To this end we shall express C_1C_2 in terms of rational numbers combined by rational operations. We shall do this in two different ways.

(a) Consider inscribed quadrilateral ABC_1C_2 in Fig. 49a. According to Ptolemy's theorem, if a quadrilateral is inscribed in a circle, the sum of the products of two pairs of opposite sides is equal to the product of the diagonals.† Therefore

$$(C_1C_2)(AB) + (AC_2)(BC_1) = (AC_1)(BC_2) \text{ or } C_1C_2 \cdot 2 + b_2a_1 = a_2b_1\,,$$

where a_i, b_i denote the legs BC_i, AC_i of $\triangle ABC_i$. It follows that C_1C_2 has the rational value

$$C_1C_2 = \tfrac{1}{2}(a_2b_1 - a_1b_2)\,.$$

(b) Place the center of the unit circle on the origin of a coordinate system so that diameter AB lies on the x-axis; denote the foot of the perpendicular from C to AB by D, see Fig. 49b. If a and b are rational, vertex C has rational coordinates (x, y) because, by similar triangles ACB, ADC, CDB,

$$\frac{1+x}{b} = \frac{b}{2}\,, \qquad \text{so} \qquad x = \frac{b^2}{2} - 1 \quad \text{is rational}\,,$$

†For a proof, see e.g. Geometry Revisited by H. S. M. Coxeter and S. L. Greitzer, NML vol. 19, p. 42. Another nice proof is based on identifying the four vertices of the inscribed quadrilateral with four complex numbers of absolute value 1.

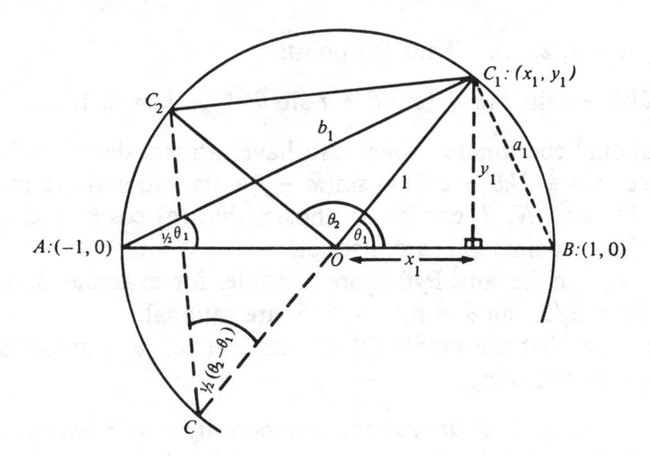

Figure 49b

and

$$\frac{y}{a} = \frac{b}{2}, \quad \text{so} \quad y = \frac{ab}{2} \quad \text{is rational}.$$

Let θ be the angle CO makes with OB; then $\angle CAB = \theta/2$, and

$$\text{(3)} \qquad \cos\frac{\theta}{2} = \frac{1+x}{b}, \qquad \sin\frac{\theta}{2} = \frac{y}{b} \quad \text{are rational}.$$

As before, let C_1 and C_2 be vertices of "rational" triangles, and let $\theta_1 = \angle C_1OB$, $\theta_2 = \angle C_2OB$. Draw diameter C_1OC_1' and chord C_2C_1', and note that $\angle C_1OC_2 = \theta_2 - \theta_1$. It follows that $\angle C_1C_1'C_2 = \frac{1}{2}(\theta_2 - \theta_1)$. Now in right $\triangle C_1C_2C_1'$,

$$C_1C_2 = 2\sin\left(\frac{\theta_2}{2} - \frac{\theta_1}{2}\right) = 2\left[\sin\frac{\theta_2}{2}\cos\frac{\theta_1}{2} - \cos\frac{\theta_2}{2}\sin\frac{\theta_1}{2}\right]$$

which, in view of (3), is rational.

Second solution. The coordinates of a point on the unit circle are $(\cos\theta, \sin\theta)$; the distance between two such points $P = (\cos\theta, \sin\theta)$ and $Q = (\cos\varphi, \sin\varphi)$ is $2|\sin\frac{1}{2}(\theta - \varphi)|$, see Figure 49b. To solve the problem, we must find 1975 angles $\theta_1, \theta_2, \ldots$ such that

$$\text{(1)} \qquad \sin\tfrac{1}{2}(\theta_k - \theta_j) \quad \text{is rational}.$$

From de Moivre's theorem:† $\cos n\theta + i\sin n\theta = (\cos\theta + i\sin\theta)^n$, it follows that if $\cos\theta$ and $\sin\theta$ are both rational, so are $\cos n\theta$ and

†See note preceding the second solution of 1962/4, p. 49

$\sin n\theta$ for $n = 1, 2, \ldots$. Thus the points

(2) $\quad \cos 2k\theta + i \sin 2k\theta = (\cos\theta + i\sin\theta)^{2k}\ , \quad k = 0, 1, \ldots, N\ ,$

all have rational coordinates. They also have rational distances from each other, since $\sin\frac{1}{2}(2k\theta - 2j\theta) = \sin(k-j)\theta$ is rational. It remains to show that for any N, θ can be so chosen that (a) $\cos\theta$ and $\sin\theta$ are rational, (b) the points (2) are distinct.

(a) Let a, b, c be any Pythagorean triple, for example 3, 4, 5; then $\cos\theta = a/c = 3/5$, $\sin\theta = b/c = 4/5$ are rational.

(b) To show that the points (2) are distinct for $\theta = \arccos(a/c)$, we shall invoke the following

LEMMA. *If $\cos\theta$, $\sin\theta$ are rational numbers a/c, b/c with values different from* 0, 1, *and* -1, *then θ is not a rational multiple of π.* (For a proof see e.g. Appendix D of I. Niven's *Numbers: Rational and Irrational*, NML vol. 1, p. 129.)

PROOF: Suppose to the contrary, that

$$\cos 2l\theta + i\sin 2l\theta = \cos 2m\theta + i\sin 2m\theta\ , \qquad l \neq m\ ,$$

i.e. that two distinct multiples of θ yield the same point on the unit circle. Then their difference $(m-l)\theta$ is a multiple of 2π, say $(m-l)\theta = 2n\pi$, and $\theta = 2n\pi/(m-l)$ would be a rational multiple of π, contrary to our lemma. We conclude that for arbitrary N, and in particular for $N = 1975$, the points (2) are all different from each other.

Remark. Observe that in our first solution, we use 1975 different primitive Pythagorean triples to solve the problem. In the second, we use only one Pythagorean triple a, b, c and prove that integer multiples of the angle $2\theta = 2\arccos(a/c)$ may be used to generate as many points as we please.

The lemma in the second solution may be avoided, for we can prove that the points (2) are distinct by showing that θ can be so chosen that $\cos\theta + i\sin\theta$ is not a $2k$-th root of 1 for any $k \leqslant N$. We merely observe that the number of $2k$-th roots of unity is $2k$, so for $k \leqslant N$, the total number of $2k$-th roots of 1 is

$$\sum_{k=1}^{N} 2k = N(N+1)\ .$$

But there are infinitely many primitive Pythagorean triples a, b, c, each corresponding to an angle θ, so that we can avoid the $N^2 + N$ angles that lead to roots of unity.

1975/6 First Solution(due to Steven Tschantz). We will show that the only continuous function $P(x, y)$ satisfying (i), (ii) and (iii) is

(1) $$P(x, y) = (x+y)^{n-1}(x-2y)\ .$$

(Thus the hypothesis that $P(x, y)$ is a polynomial is actually superfluous.)

In (ii) put $b = 1 - a$ and $c = 0$. We get

$$P(1-a,a) + P(a,1-a) + P(1,0) = 0 \ .$$

Since $P(1,0) = 1$, this yields

$$P(1-a,a) = -1 - P(a,1-a) \ . \tag{2}$$

Next set $c = 1-a-b$ in (ii). We obtain

$$P(1-a,a) + P(1-b,b) + P(a+b,1-a-b) = 0 \ .$$

By (2), this implies that

$$-2 - P(a,1-a) - P(b,1-b) + P(a+b,1-a-b) = 0 \ ,$$

or equivalently

$$P(a+b,1-a-b) = P(a,1-a) + P(b,1-b) + 2 \ . \tag{3}$$

Now let $f(x) = P(x,1-x) + 2$. Equation (3) then takes the form

$$f(a+b) - 2 = f(a) - 2 + f(b) - 2 + 2 \ ,$$

or in other words

$$f(a+b) = f(a) + f(b) \ . \tag{4}$$

Since $P(x,y)$ is continuous, so is $f(x)$.

A famous theorem of Cauchy asserts that *the only continuous functions* $f(x)$ *satisfying* (4) *are* $f(x) = kx$, *where* k *is a constant*. In order not to interrupt the discussion, we postpone the proof of this result to the end of this solution and apply it in the present case. To evaluate the constant k, we note that

$$f(1) = P(1, 0) + 2 = 1 + 2 = 3 \ .$$

Therefore $k = 3$, so $f(x) = 3x$. Since by definition

$$f(x) = P(x, 1-x) + 2,$$

it follows that

$$P(x, 1-x) = 3x - 2 \ . \tag{5}$$

Now if $a + b \neq 0$, we can set $t = a + b$, $x = \dfrac{a}{a+b}, y = \dfrac{b}{a+b}$ in (i). This yields

$$P(a, b) = (a+b)^n P\left(\frac{a}{a+b}, \frac{b}{a+b}\right) . \tag{6}$$

In equation (5), take $x = a/(a+b)$. Then $1 - x = b/(a+b)$, so we get

$$P\left(\frac{a}{a+b}, \frac{b}{a+b}\right) = \frac{3a}{a+b} - 2 = \frac{a-2b}{a+b} \ .$$

Substituting this into (6), we find that

$$P(a, b) = (a + b)^n \frac{a - 2b}{a + b} = (a + b)^{n-1}(a - 2b)$$

whenever $a + b \neq 0$. Since $P(x, y)$ is continuous, it follows that the identity $P(a, b) = (a + b)^{n-1}(a - 2b)$ continues to hold even when $a + b = 0$. Thus the only continuous function which can satisfy the given conditions is the polynomial $P(x, y)$ defined in (1). It is straightforward to verify conversely that this polynomial satisfies (i), (ii) and (iii).

We now turn to the proof of the theorem of Cauchy which was used above.

THEOREM. *Let $f(x)$ be a continuous real function of a real variable x. Suppose that f satisfies the functional equation $f(a + b) = f(a) + f(b)$ for all a and b. Then $f(x) = kx$, where $k = f(1)$.*

PROOF: We first show that

(7) $$f(nx) = nf(x) \text{ for all positive integers } n\ .$$

The proof is by mathematical induction on n. Clearly (7) holds when $n = 1$. Suppose it holds for some integer $n \geqslant 1$. Then, using the functional equation $f(a + b) = f(a) + f(b)$, we find that

$$\begin{aligned} f((n + 1)x) &= f(nx + x) = f(nx) + f(x) \\ &= nf(x) + f(x) = (n + 1)f(x)\ . \end{aligned}$$

Thus (7) also holds for $n + 1$, completing the induction.

Now let $k = f(1)$. If q is any positive integer, it follows from (7) with $n = q$, $x = 1/q$, that $f(1) = qf(1/q)$. Hence $f(1/q) = k/q$. Next suppose p is any positive integer, and apply (7) with $n = p$, $x = 1/q$. We obtain

$$f\left(\frac{p}{q}\right) = pf\left(\frac{1}{q}\right) = p\frac{k}{q} = k\frac{p}{q}\ .$$

In other words, the equation $f(x) = kx$ holds whenever x is a positive rational number p/q. Since the rational numbers are everywhere dense on the real line, and since f is continuous, it follows that $f(x) = kx$ for all non-negative real numbers x. In particular $f(0) = 0$. Hence $f(x) + f(-x) = f(x - x) = f(0) = 0$, so $f(-x) = -f(x)$. Therefore the identity $f(x) = kx$ also holds for negative values of x.

Second solution. We now give a less elegant, but perhaps more systematic solution using the hypothesis that $P(x, y)$ is a polynomial. If a, b, c are all equal, condition (ii) yields

$$3P(2a, a) = 0 \qquad \text{for all } a\ ,$$

so that $P(x, y)$ vanishes for $x - 2y = 0$. It is not difficult to prove (and we leave this to the reader) that P has $(x - 2y)$ as a factor, that is

(8) $$P(x, y) = (x - 2y)Q(x, y)\ ,$$

where Q is a homogeneous polynomial of degree $n - 1$. We note that

$Q(1, 0) = P(1, 0) = 1$. Condition (ii) with $b = c$ says

$$P(2b, a) + 2P(a + b, b) = 0.$$

We write this in terms of Q as defined in (8) and get

$$(2b - 2a)Q(2b, a) + 2(a - b)Q(a + b, b) = 2(a - b)[Q(a + b, b) - Q(2b, a)] = 0 .$$

Hence

$$(9) \qquad Q(a + b, b) = Q(2b, a) \qquad \text{whenever } a \neq b .$$

Obviously (9) holds also when $a = b$. With $a + b = x$, $b = y$ so that $a = x - y$, (9) becomes

$$Q(x, y) = Q(2y, x - y) .$$

This functional equation states that replacing the first and second argument of Q by twice the second and the first minus the second, respectively, does not alter the value of Q. Repeating this principle leads to

$$(10) \qquad \begin{aligned} Q(x, y) &= Q(2y, x - y) = Q(2x - 2y, 3y - x) \\ &= Q(6y - 2x, 3x - 5y) = \dots, \end{aligned}$$

where the sum of the arguments is always $x + y$. Each member of (10) may be written as

$$Q(x, y) = Q(x + d, y - d)$$

with

$$(11) \qquad d = 0, 2y - x, x - 2y, 6y - 3x, \cdots .$$

It is easily seen that these values of d are all distinct if $x \neq 2y$. For any fixed values of x and y, the equation

$$Q(x + d, y - d) - Q(x, y) = 0$$

is a polynomial equation of degree $n - 1$ in d, and if $x \neq 2y$ it has infinitely many solutions (some of which are given in (11)). It follows that if $x \neq 2y$, the equation

$$Q(x + d, y - d) = Q(x, y)$$

holds for all d. By continuity it holds also if $x = 2y$. But this means that $Q(x, y)$ is a function of the single variable $x + y$. Since it is homogeneous of degree $n - 1$, we have

$$Q(x, y) = c(x + y)^{n-1} , \quad \text{where } c \text{ is a constant} .$$

Since $Q(1, 0) = 1$, we have $c = 1$, and hence

$$P(x, y) = (x - 2y)(x + y)^{n-1} .$$

Eighteenth International Olympiad, 1976

1976/1. Let $ABCD$ be a quadrilateral with

$$AB + BD + DC = 16 \ . \tag{1}$$

Its area K, which we express as

$$K = \tfrac{1}{2}AB \cdot BD \sin \angle ABD + \tfrac{1}{2}DC \cdot BD \sin \angle CDB \tag{2}$$

depends on angles ABD and CDB. To determine the possible lengths of AC for $K = 32$, let us first determine the maximum value of K under the given condition (1).

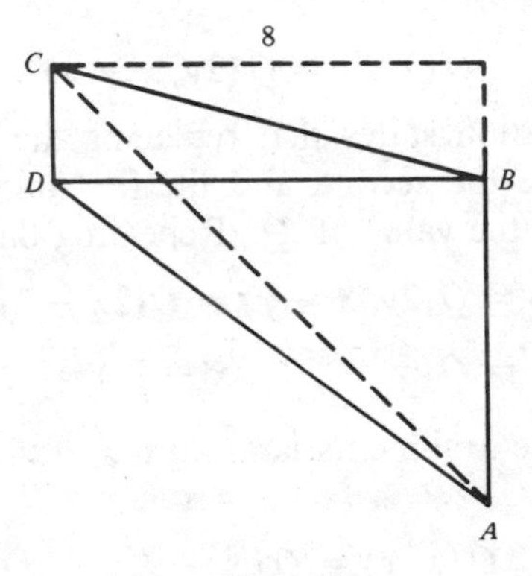

Figure 50

No matter what the separate lengths of AB, BD and DC are, K is certainly largest when $\angle ABD = \angle CDB = 90°$, see Fig. 50, so that maximizing the area (2) under condition (1) is equivalent to maximizing

$$BD(AB + DC) \text{ subject to } AB + BD + DC = 16 \ . \tag{3}$$

By the arithmetic-mean-geometric-mean inequality,† it follows that the product of two positive numbers (BD and $AB + DC$ in our case) whose sum is constant is largest when these numbers are equal; that is, when $BD = AB + DC$. But then, by (1)

$$BD = AB + DC = 8 \ , \tag{4}$$

and by (2)

$$K = \tfrac{1}{2}BD(AB + DC) = 4 \cdot 8 = 32 \ .$$

Thus the conditions of the problem are met only if (4) holds, and the

†See p. 196.

diagonal AC, in this case, has length

$$AC = \sqrt{(AB + DC)^2 + BD^2} = \sqrt{128} = 8\sqrt{2} ,$$

calculated by the Pythagorean theorem, see Fig. 50.

1976/2 First solution. The polynomials

$$P_n(x) = \left[\cdots \left((x^2 - 2)^2 - 2\right)^2 - \cdots \right]^2 - 2$$

are of degree 2^n and are even functions, that is,

$$P_n(-x) = P_n(x) .$$

We observe that P_1 maps the interval $[-2, 2]$ onto the interval $[-2, 2]$. As x increases from -2 to 0, P_1 decreases from 2 to -2, and since P_1 is even, as x increases from 0 to 2, P_1 increases from -2 to 2; see Fig. 51. Next consider $P_2(x) = P_1(P_1)$. As x goes from -2 to 0, the output of P_1, from 2 to -2, is fed again into P_1, which maps it onto $[-2, 2]$, see the interval $-2 \leqslant x \leqslant 0$ in Fig. 51. (The graph for $0 \leqslant x \leqslant 2$ is just the mirror image.) By the same reasoning, the graph of P_3 goes from the ordinate 2 to the ordinate -2 twice as often as P_2 did, etc. Thus $P_n(x)$ covers $(-2, 2)$ 2^n times. The solutions of $P_n(x) = x$ are the abscissas of the points where the graph of P_n intersects the graph of $y = x$. This line, for $-2 < x < 2$, is just the diagonal of the 4×4 square in which P_n travels down and up 2^n times. Hence there are 2^n distinct intersections furnishing 2^n real distinct roots of $P_n(x) = x$. Since $P_n(x)$ is of degree 2^n, $P_n(x) = x$ cannot have more than 2^n roots; so all its roots are real and distinct.

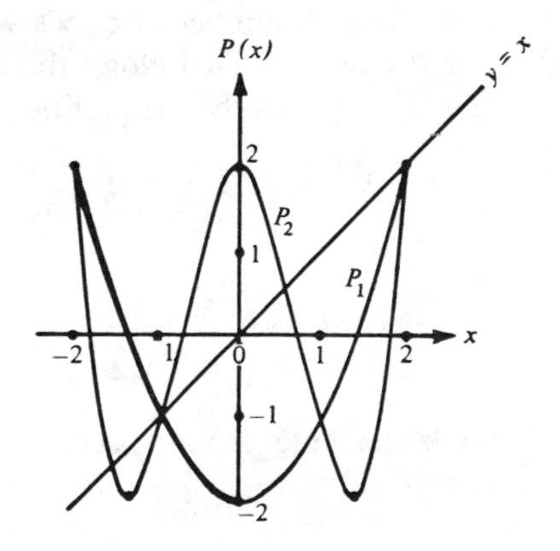

Figure 51

Second solution. We set $x(t) = 2\cos t$ and observe that this function maps $0 \leqslant t \leqslant \pi$ into $2 \geqslant x \geqslant -2$. With the help of double angle formulas, we find that

$$P_1(x(t)) = P_1(2\cos t) = 4\cos^2 t - 2 = 2\cos 2t$$

$$P_2(x(t)) = P_1(P_1(x(t))= 4\cos^2 2t - 2 = 2\cos 4t$$

. . . .

and, in general,

$$P_n(x(t)) = 2\cos 2^n t\ .$$

The equation $P_n(x) = x$ is transformed into

$$2\cos 2^n t = 2\cos t$$

and is satisfied whenever $2^n t = \pm t + 2k\pi$, $k = 0, 1, \ldots$, i.e. when

$$t = \frac{2k\pi}{2^n - 1} \quad \text{and} \quad t = \frac{2k\pi}{2^n + 1}\ .$$

The first expression yields 2^{n-1} distinct values of $\cos t$ for $k = 0, 1, \ldots, 2^{n-1} - 1$, the second yields another 2^{n-1} distinct values of $\cos t$ for $k = 1, 2, \ldots, 2^{n-1}$. It follows that there are altogether $2 \times 2^{n-1} = 2^n$ real distinct values of $x = 2\cos t$ satisfying the equation $P_n(x) = x$.

Note. The n-th Chebyshev polynomial $T_n(x)$ is defined as $\cos n\theta$ with $x = \cos\theta$. It is easy to verify that the $P_n(x)$ of our problem are the Chebyshev polynomials $T_{2^n}(2x)$. Chebyshev polynomials play important parts in the approximation of functions, of integrals, and in interpolation; see also Remark 3 following the solution of 1972/5.

1976/3. Since unit cubes fill the box completely, each of its dimensions is a natural number, say a_1, a_2 and a_3. The volume of the box, then, is $a_1a_2a_3$. Now let b_i be the largest number of cubes with volume 2, hence with edge length $2^{1/3}$, that can be placed along the edge of length a_i of our box. Thus the integer b_i satisfies the inequalities

$$a_i - \sqrt[3]{2} < \sqrt[3]{2}\, b_i \leqslant a_i\ ,$$

i.e.

$$\frac{a_i}{\sqrt[3]{2}} - 1 < b_i \leqslant \frac{a_i}{\sqrt[3]{2}}\ . \tag{1}$$

In other words, b_i is the largest integer not exceeding $a_i/\sqrt[3]{2}$, and this is usually denoted by brackets:

$$b_i = [a_i/\sqrt[3]{2}\,]\ . \tag{2}$$

The volume occupied by these cubes is $\sqrt[3]{2}\,b_1 \cdot \sqrt[3]{2}\,b_2 \cdot \sqrt[3]{2}\,b_3$, and this, we are told, is 40% of the box volume. Thus $2b_1b_2b_3 = (2/5)a_1a_2a_3$, whence

$$\frac{a_1a_2a_3}{b_1b_2b_3} = 5\,. \tag{3}$$

To find all positive integers a_1, a_2, a_3 subject to conditions (2) and (3), we observe that $a_i > 1$ (otherwise $b_i = 0$) and tabulate a few values of a, $b = [a/\sqrt[3]{2}\,]$, and a/b:

a	2	3	4	5	6	7	8	$\cdots$
b	1	2	3	3	4	5	6	$\cdots$
a/b	2	1.5	$1.33\cdots$	$1.66\cdots$	1.5	1.4	$1.33\cdots$	$\cdots$

We claim that for $a > 8$, $a/b < 1.5$. To prove this we need only use the estimate $b > (a/\sqrt[3]{2}\,) - 1$, see (1), from which it follows that

$$\frac{a}{b} < \frac{a}{\dfrac{a}{\sqrt[3]{2}} - 1} = \frac{\sqrt[3]{2}}{1 - \dfrac{\sqrt[3]{2}}{a}}\,.$$

As a increases, the denominator in the last member increases, hence the fraction decreases.

$$\text{For } a \geqslant 8\,, \quad \frac{a}{b} < \frac{\sqrt[3]{2}}{1 - \dfrac{\sqrt[3]{2}}{8}} < \frac{1.26}{1 - \dfrac{1.26}{8}} < 1.5\,. \tag{4}$$

This together with the table above yields that

$$\text{for } a \geqslant 3\,, \quad \frac{a}{b} \leqslant \frac{5}{3}\,. \tag{5}$$

If a_1, a_2, a_3 were at least 3, then $a_1a_2a_3/b_1b_2b_3 \leqslant (5/3)^3$ would violate (3); so we conclude that at least one of the a's, say a_1, must be 2. Consequently by (3)

$$\frac{a_2}{b_2}\,\frac{a_3}{b_3} = \frac{5}{2}\,. \tag{6}$$

Next we show that both other dimensions of our box are greater than 2. We accomplish this by demonstrating that for $i = 2, 3$, $a_i/b_i < 2$, from which $a_i > 2$ will follow. By (1), $b_i \leqslant a_i/\sqrt[3]{2}$, so $b_i/a_i \leqslant 1/\sqrt[3]{2}$; and by (6), with $i = 3$,

$$\frac{a_2}{b_2} = \frac{5}{2}\,\frac{b_3}{a_3} \leqslant \frac{5}{2\sqrt[3]{2}} < 2\,.$$

Similarly, with $i = 2$, (1) and (6) yield $a_3/b_3 < 2$.

Finally, we note from the table and (4) that $a/b \leqslant 3/2$ unless $a = 2$ or $a = 5$. Since $(3/2)^2 = 9/4 < 5/2$, we see that (6) would be violated unless one of a_2, a_3, say a_2, is 5. Then $a_2/b_2 = 5/3$ and (6) yields

$$\frac{5}{3}\cdot\frac{a_3}{b_3} = \frac{5}{2}\,, \qquad \frac{a_3}{b_3} = \frac{3}{2}\,.$$

Thus a_3 is either 3 or 6. So the dimensions of the box are either $2 \times 3 \times 5$ or $2 \times 5 \times 6$.

1976/4. The number of different partitions of 1976 into positive integers is finite. Therefore, there is a finite number of associated products among which there must be a largest. So the problem has a solution.

Suppose we have a partiton such that

$$a_1 + a_2 + \cdots + a_n = 1976\,;$$

the associated product is $\prod_{i=1}^{n} a_i = a_1 \cdot a_2 \cdot \cdots \cdot a_n$. We shall replace some of the a_i in such a way that the sum does not change, but the product increases.

If the partition contains an $a_j \geqslant 4$, we replace it by the two numbers 2 and $a_j - 2$. The sum is not affected since $2 + (a_j - 2) = a_j$, but the factor a_j in the product is replaced by

$$2(a_j - 2) = 2a_j - 4\,,$$

and if $a_j \geqslant 4$, then $2a_j \geqslant 4 + a_j$, so $2a_j - 4 \geqslant a_j$. This process transforms numbers > 3 into 2's and 3's, and the new partition has an associated product at least as large as that of the old. Furthermore, if $a_i = 1$, add it to some a_j, and replace the summands $1, a_j$ by the single number $a_j + 1$. The sum remains the same while the factors $1 \times a_j = a_j$ are replaced by the factor $a_j + 1$, leading to a larger associated product. This process leads to the creation of numbers > 3 only if $a_j = 3$; but then $a_j + 1 = 4$ can be broken up into 2 and 2 without changing sum or product, as explained above. After getting rid of all 1's, we shall be left with only 2's and 3's in our partition. The associated product is of the form

$$2^x \cdot 3^y\,.$$

If $x \geqslant 3$, we replace each triple of 2's by a pair of 3's. Since $2 + 2 + 2 = 3 + 3$, this does not change the sum, but since $2^3 < 3^2$, this further increases the product. Thus the largest product has the form

$$2^a \cdot 3^b\,, \qquad a = 0, 1, \text{or } 2\,.$$

Since $1976 = 3 \times 658 + 2$, its "best" partition consists of 658 3's and one 2, and the largest product is the number 2×3^{658}.

1976/5. Let $(y_1, y_2, \cdots, y_q)$ be a q-tuple of integers such that $|y_j| \leqslant p$, $j = 1, 2, \cdots, q$. Then the value of the left member of the r-th equation is some integer between $-pq$ and pq, since the coefficients are $-1, 0$ or 1. Thus the left member

$$\sum_{i=1}^{q} a_{ri} y_i$$

can have at most $2pq + 1$ values: pq positive integer values, pq negative integer values and the value 0. Now consider the p-tuple of all p left sides in our system. Since each can take at most $2pq + 1$ values, at most $(2pq + 1)^p$ distinct p-tuples can result. Each y_j is an integer between $-p$ and p, so there are $2p + 1$ choices for each y_j, and since there are q y's in a q-tuple, we can make up a total $(2p + 1)^q$ different ordered q-tuples.

Now $q = 2p$, so the number of q-tuples $(y_1, \cdots, y_q)$ with $|y_j| \leqslant p$ is

$$(2p + 1)^q = (2p + 1)^{2p} = \left[(2p + 1)^2\right]^p = \left[4p^2 + 4p + 1\right]^p ,$$

while the number of p-tuples

$$\left(\sum_{j=1}^{q} a_{1j} y_j, \ \sum_{j=1}^{q} a_{2j} y_j \ \cdots, \ \sum_{j=1}^{q} a_{pj} y_j \right)$$

they can generate is at most

$$(2pq + 1)^p = (4p^2 + 1)^p .$$

Therefore there are more q-tuples $(y_1, \cdots, y_q)$ than there are value-sets, and by the pigeonhole principle†, there are at least two distinct q-tuples producing the same values of the left sides. Denote these q-tuples by

$$(1) \qquad (y_1, y_2, \cdots, y_q) \quad \text{and} \quad (z_1, z_2, \cdots, z_q) .$$

We claim that the q-tuple $(x_1, x_2, \cdots, x_q)$ of differences $y_j - z_j = x_j$, $j = 1, 2, \cdots, q$, is a solution of the problem satisfying properties (a), (b), (c). To verify this claim, we first observe that

$$\sum_{j=1}^{q} a_{rj} y_j = \sum_{j=1}^{q} a_{rj} z_j , \qquad r = 1, 2, \cdots p$$

implies

$$\sum_{j=1}^{q} a_{rj} x_j = \sum_{j=1}^{q} a_{rj} (y_j - z_j) = \sum_{j=1}^{q} a_{rj} y_j - \sum_{j=1}^{q} a_{rj} z_j = 0 .$$

†See p. 201.

So the x_i satisfy all p equations. Moreover, since y_i and z_i are integers, so are their differences, and (a) is satisfied. The q-tuples (1) are distinct, so not all x_j are zero; thus (b) is satisfied. Finally, since $|y_j| \leq p$ and $|z_j| \leq p$, we see by the triangle inequality that $|x_j| = |y_j - z_j| \leq |y_j| + |z_j| \leq 2p$, so $|x_j| \leq q$; (c) also is satisfied.

1976/6. In an attempt to discover a pattern, we calculate a few members of the sequence:

$$u_2 = \frac{5}{2}(2^2 - 2) - \frac{5}{2} = \frac{5}{2}, \qquad u_3 = \frac{5}{2}\left[\left(\frac{5}{2}\right)^2 - 2\right] - \frac{5}{2} = \frac{65}{8}$$

$$u_4 = \frac{65}{8}\left[\left(\frac{5}{2}\right)^2 - 2\right] - \frac{5}{2} = \frac{1025}{32}, \qquad u_5 = \frac{1025}{32}\left[\left(\frac{65}{8}\right)^2 - 2\right] - \frac{5}{2}$$

$$= \frac{4194305}{2048}.$$

We observe that the denominators are powers of 2, and the numerators are 1 plus a power of 2:

$$u_0 = \frac{2^0 + 1}{2^0} = 2^0 + 2^{-0}, \qquad u_1 = \frac{2^2 + 1}{2^1} = 2^1 + 2^{-1} = u_2,$$

$$u_3 = \frac{2^6 + 1}{2^3} = 2^3 + 2^{-3}, \qquad u_4 = \frac{2^{10} + 1}{2^5} = 2^5 + 2^{-5},$$

$$u_5 = 2^{11} + 2^{-11}, \ldots.$$

We suspect that all have the form

(1) $$u_n = 2^{f(n)} + 2^{-f(n)};$$

and since, for $n > 0$, the term $2^{-f(n)}$ seems to be a proper fraction, it will not contribute to $[u_n]$. We therefore suspect that $f(n)$ may be the exponent given in the statement of the problem and proceed to verify this suspicion. It is true for $n = 1, 2, 3$.

Set

(2) $$f(n) = \tfrac{1}{3}[2^n - (-1)^n].$$

For $n = 1, 2, \ldots$, $f(n)$ is clearly positive; it is, in fact, a natural number, because $2 \equiv -1 \pmod 3$, so $2^n \equiv (-1)^n \pmod 3$, whence $2^n - (-1)^n \equiv 0 \pmod 3$, so that $2^n - (-1)^n$ is divisible by 3. Moreover, our impression that $2^{-f(n)} < 1$ is now confirmed for $f(n)$ given by (2).

Next, we suppose that (1) holds for all $k \leqslant n$, and substitute this into the recursion formula. We get

$$\begin{aligned} u_{n+1} &= u_n\left(u_{n-1}^2 - 2\right) - \tfrac{5}{2} \\ &\doteq (2^{f(n)} + 2^{-f(n)})(2^{2f(n-1)} + 2^{-2f(n-1)}) - \tfrac{5}{2} \\ &= 2^{f(n)+2f(n-1)} + 2^{-f(n)-2f(n-1)} + 2^{f(n)-2f(n-1)} \\ &\qquad + 2^{-f(n)+2f(n-1)} - \tfrac{5}{2}\ . \end{aligned}$$

Now

$$\begin{aligned} f(n) + 2f(n-1) &= \tfrac{1}{3}2^n + \tfrac{2}{3}2^{n-1} - \tfrac{1}{3}(-1)^n \quad - \tfrac{2}{3}(-1)^{n-1} \\ &= \tfrac{1}{3}2^{n+1} \qquad\quad + \tfrac{1}{3}(-1)^{n+1} - \tfrac{2}{3}(-1)^{n+1} \\ &= \tfrac{1}{3}[2^{n+1} - (-1)^{n+1}] = f(n+1)\ , \end{aligned}$$

and

$$\begin{aligned} f(n) - 2f(n-1) &= \tfrac{1}{3}2^n - \tfrac{2}{3}2^{n-1} - \tfrac{1}{3}(-1)^n + \tfrac{2}{3}(-1)^{n-1} \\ &= \tfrac{1}{3}(2^n - 2^n) + \tfrac{1}{3}(-1)^{n+1} + \tfrac{2}{3}(-1)^{n+1} \\ &= \quad 0 \quad + \quad (-1)^{n+1} = (-1)^{n+1}\ . \end{aligned}$$

Therefore, writing $5/2 = 2 + 2^{-1}$, we see that

$$\begin{aligned} u_{n+1} &= 2^{f(n+1)} + 2^{-f(n+1)} + 2 + 2^{-1} - \tfrac{5}{2} \\ &= 2^{f(n+1)} + 2^{-f(n+1)}\ , \end{aligned}$$

completing the induction. In particular, $[u_{n+1}] = 2^{f(n+1)}$ for all positive integers n.

Nineteenth International Olympiad, 1977

1977/1 First solution. In Figure 52, O is the center of the square. Diagonals AC and BD are axes of symmetry, and so are the lines through M, K and through L, N. Rotation about O through any multiple of 90° also leaves the figure invariant. It is therefore enough to work in the portion of the plane bounded by rays OA and OL. Subsequent reflections in OA, then OM, and finally LN will generate the rest of the figure. We have denoted the midpoints of $AK, LM, AN, \cdots$ by $P_1, P_2, P_3 \cdots$ respectively.

Since $AK = AD$ and $\angle DAK = 90° - 60° = 30°$, we have $\angle ADK = \angle AKD = 75°$. Therefore the congruent isosceles triangles CDK,

BCN, *ABM*, *DAL* have base angles of $75° - 60° = 15°$.† It follows that congruent triangles *AML*, *BNM*, *CKN* and *DLK* are equilateral. Denote the length of their sides by s.

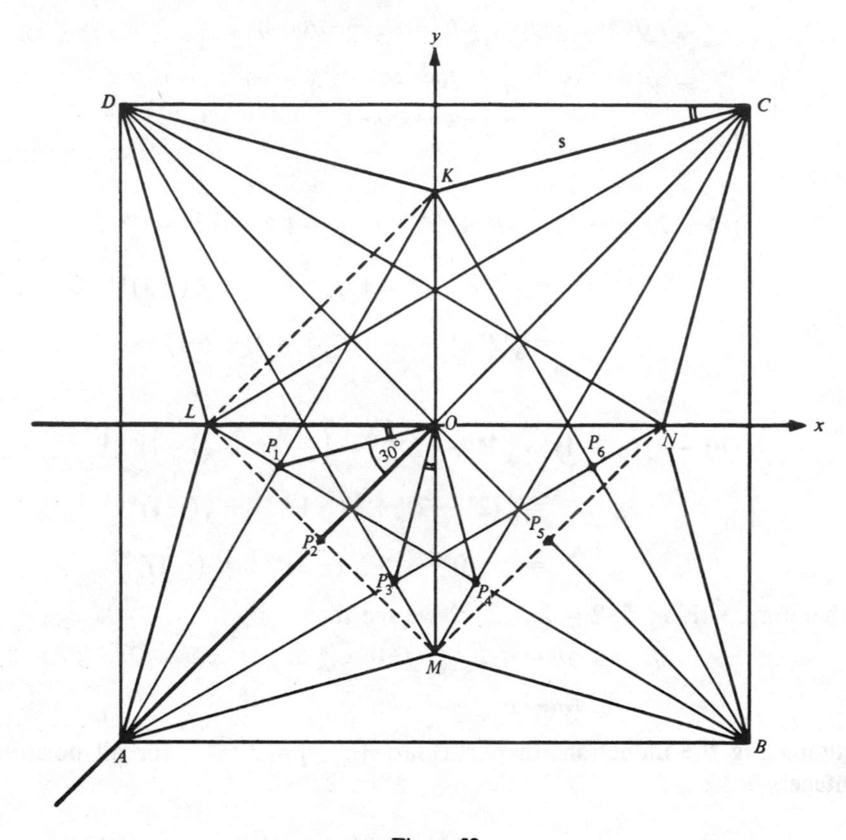

Figure 52

Segment OP_1 connecting the midpoints of sides AK and AC in $\triangle AKC$ is $\| KC$ and has length $\frac{1}{2} KC = \frac{1}{2} s$. By symmetry, $OL \| DC$, so $\angle LOP_1 = \angle DCK = 15°$. $\angle AOL = 45°$, so $\angle P_1OP_2 = 30°$. Segments AO and BO are perpendicular; BO bisects $\angle MBN$ and so is $\perp$ to MN. Hence $MN \| AO$ and $OP_2 = MP_5 = \frac{1}{2} s$; that is $OP_1 = OP_2$.

By reflection in line OA, we now have $OP_3 = OP_1$, $\angle P_2OP_3 = 30°$, $\angle P_3OM = 15°$. Next, reflect in line OM and add OP_4, OP_5 and OP_6 to the list of equal segments, and $\angle P_3OP_4 (= 2 \cdot 15°) = \angle P_4OP_5 = \angle P_5OP_6$ to the list of 30° angles.

†For more on this interesting configuration, see *Geometry Revisited* by H. S. M. Coxeter and S. L. Greitzer, NML vol. 19, 1967, p. 25.

Finally, reflection about LN produces the rest of the figure, and we have twelve points $P_1, P_2, \ldots P_{12}$ lying on a circle of radius $s/2$ and $30°$ apart on this circle. They are the twelve vertices of a regular dodecagon.

Second solution. Place the center O of the square at the origin of a coordinate plane so that the vertices of the square are the points

$$A = (-2, -2)\ , \quad B = (2, -2)\ , \quad C = (2, 2)\ , \quad D = (-2, 2)\ .$$

Note that B, C, D may be obtained from A by successive rotations through $90°$ counterclockwise about the origin.

The altitude of equilateral $\triangle ABK$ with side length 4 has length $2\sqrt{3}$, so the coordinates of point K are $K = (0, 2\sqrt{3} - 2)$, see Figure 52. Successive rotations through $90°$ about O take point K into L, M, N; thus

$$K = (0, 2\sqrt{3} - 2)\ , \quad L = (2 - 2\sqrt{3}\,, 0)\ ,$$

$$M = (0, 2 - 2\sqrt{3}\,)\ , \quad N = (2\sqrt{3} - 2, 0)\ .$$

We calculate the coordinates of the midpoints P_1, P_2, P_3 of segments AK, LM, AN, respectively, using the averages of the coordinates of the endpoints. We obtain

$$P_1 = (-1, \sqrt{3} - 2)\ , \quad P_2 = (1 - \sqrt{3}\,, 1 - \sqrt{3}\,)\ , \quad P_3 = (\sqrt{3} - 2, -1)\ .$$

Now $\sqrt{3} - 2 < 0$ and $1 - \sqrt{3} < 0$, so P_1, P_2, P_3 are in the third quadrant. The other 9 midpoints can be obtained from P_1, P_2, P_3 by successive rotations through $90°$ about O. To prove that the twelve midpoints form a regular dodecagon, it suffices, thanks to the symmetry, to show that P_1, P_2, P_3 are equally distant from O and that sides P_1P_2, P_2P_3 and P_3P_4 of the dodecagon have the same length; here P_4 is the image of P_1 under rotation by $90°$ and hence has coordinates $P_4 = (2 - \sqrt{3}\,, -1)$. (The equality $P_1P_2 = P_2P_3$ follows from the symmetry with respect to the diagonal AC of the square.) We achieve this by using the distance formula:

$$OP_1^2 = 1 + (\sqrt{3} - 2)^2 = 8 - 4\sqrt{3}\ , \quad OP_2^2 = 2(1 - \sqrt{3}\,)^2 = 8 - 4\sqrt{3}\ ,$$

$$OP_3^2 = (\sqrt{3} - 2)^2 + 1 = 8 - 4\sqrt{3}\ ,$$

while

$$P_1P_2^2 = (\sqrt{3} - 2)^2 + (2\sqrt{3} - 3)^2 = 28 - 16\sqrt{3}\ ,$$

$$P_3P_4^2 = (2\sqrt{3} - 4)^2 = 28 - 16\sqrt{3}\ .$$

This shows that all midpoints have distance $2\sqrt{2 - \sqrt{3}}$ from O and all sides of the dodecagon have length $2\sqrt{7 - 4\sqrt{3}}$.

Remark. An alternative proof consists of subjecting the vector P_1 to three successive rotations of 30° for example by multiplying it by the rotation matrix

$$R = \begin{pmatrix} \cos 30° & -\sin 30° \\ \sin 30° & \cos 30° \end{pmatrix} = \begin{pmatrix} \sqrt{3}/2 & -1/2 \\ 1/2 & \sqrt{3}/2 \end{pmatrix}$$

and then verifying that

$$RP_1 = P_2\,, \quad R^2P_1 = P_3 \quad \text{and} \quad R^3P_1 = P_4\,.$$

1977/2. Denote by S_n a sequence of n terms having the desired property. We shall prove that there is no S_{17}. Then we shall construct an S_{16} which automatically furnishes also S_n for $n < 16$. This will demonstrate that 16 is the desired maximum number of terms.

We denote the terms of S_{17} by $a_1, a_2, \ldots, a_{17}$ and write all sets of seven consecutive terms in rows of the table below:

$$\begin{array}{ccccccc} a_1 & a_2 & a_3 & a_4 & a_5 & a_6 & a_7 \\ a_2 & a_3 & a_4 & a_5 & a_6 & a_7 & a_8 \\ a_3 & a_4 & a_5 & a_6 & a_7 & a_8 & a_9 \\ \vdots & \vdots & \vdots & \vdots & \vdots & \vdots & \vdots \\ a_{11} & a_{12} & a_{13} & a_{14} & a_{15} & a_{16} & a_{17}\,. \end{array}$$

We note that the columns in this table contain all sets of eleven consecutive terms of S_{17}. Now we add all entries of the table first by adding all entries in each row, then by adding the row sums. By hypothesis, each row sum is negative, hence the total is also negative. Then we add all entries of the table by adding the entries in each column and then adding the column sums. By hypothesis each column sum is positive, hence the total is positive, which contradicts what we just found by adding the rows. We conclude that there is no S_{17}.

An example of S_{16} is the sequence

$$5, 5, -13, 5, 5, 5, -13, 5, 5, -13, 5, 5, 5, -13, 5, 5,$$

and any n consecutive terms of S_{16} lead to a sequence S_n for $11 \leqslant n \leqslant 16$. Thus to obtain an S_{14}, just pick any 14 consecutive terms of S_{16}.

We shall now show how the above S_{16} can be found by a method other than guessing.

Assume we can find a sequence S_{16} that reads the same from left to right as from right to left, and that the sum of any seven consecutive terms is -1 and the sum of any eleven consecutive terms is 1. Then

$$a_1 + a_2 + a_3 + a_4 + a_5 + a_6 + a_7 = -1$$
$$a_2 + a_3 + a_4 + a_5 + a_6 + a_7 + a_8 = -1$$
$$a_3 + a_4 + a_5 + a_6 + a_7 + a_8 + a_8 = -1$$
$$a_4 + a_5 + a_6 + a_7 + a_8 + a_8 + a_7 = -1$$
$$a_5 + a_6 + a_7 + a_8 + a_8 + a_7 + a_6 = -1 .$$

Subtracting the second equation from the first, then the third from the second, etc., we obtain

$$a_1 = a_8 , \qquad a_2 = a_8 , \qquad a_3 = a_7 , \qquad a_4 = a_6 .$$

Now writing the sum of eleven consecutive terms

$$a_1 + a_2 + a_3 + a_4 + a_5 + a_6 + a_7 + a_8 + a_8 + a_7 + a_6 = 1$$
$$a_2 + a_3 + a_4 + a_5 + a_6 + a_7 + a_8 + a_8 + a_7 + a_6 + a_5 = 1$$
$$a_3 + a_4 + a_5 + a_6 + a_7 + a_8 + a_8 + a_7 + a_6 + a_5 + a_4 = 1$$

and subtracting the second equation from the first and the third from the second, we obtain

$$a_1 = a_5 \qquad \text{and} \qquad a_2 = a_4 .$$

It follows that S_{16} has the form

$$a_1, a_1, a_3, a_1, a_1, a_1, a_3, a_1, a_1, a_3, a_1, a_1, a_1, a_3, a_1, a_1 ,$$

where the sum of any seven consecutive terms is $5a_1 + 2a_3 = -1$, and the sum of any eleven consecutive terms is $8a_1 + 3a_3 = 1$. The solution of this pair of equations is $a_1 = 5$, $a_3 = -13$ and leads to the S_{16} above.

1977/3. The set V_n is a multiplicative system in the sense that its elements, characterized by

$$v \equiv 1 \pmod{n} \qquad \text{and} \qquad v > 1$$

have the property that the product of any two of them is also in V_n. Now indecomposable elements of V_n are analogues of primes in the multiplicative system of ordinary natural numbers. The same method used to prove that every natural number can be factored into primes serves to prove that every element of V_n can be factored into indecomposables. The point of this problem is to show that, in contrast to the situation in the set of natural numbers, factorization into indecomposables in V_n is *not* unique

Let $a = n - 1$, $b = 2n - 1$ and define r by

$$r = (a^2)(b^2) = (ab)(ab) . \tag{1}$$

Neither a nor b belongs to V_n since $a \equiv b \equiv -1 \pmod{n}$. But a^2, ab, and b^2 are in V_n since $a^2 \equiv ab \equiv b^2 \equiv 1 \pmod{n}$.

We claim that a^2 is indecomposable in V_n. For if it were decomposable, then

$$a^2 = (cn + 1)(dn + 1) \quad \text{for some natural numbers } c, d \;,$$

and we would have

$$a^2 = cdn^2 + (c + d)n + 1 > n^2 = (a + 1)^2 \;,$$

a contradiction. Furthermore the indecomposable number a^2 is not a divisor of ab because

$$\frac{ab}{a^2} = \frac{b}{a} = \frac{2n - 1}{n - 1} = 2 + \frac{1}{n - 1} \;,$$

and this is not an integer for $n > 2$.

Now equation (1) leads to two distinct factorings of r into indecomposables, because whether or not b^2 and ab are decomposable the factor a^2 cannot appear among the factors of ab.

1977/4. Let $\sqrt{a^2 + b^2} = r$. Then $(a/r)^2 + (b/r)^2 = 1$, and there is an angle α such that

$$\frac{a}{r} = \cos \alpha \;, \qquad \frac{b}{r} = \sin \alpha \;.$$

We use these expressions to write

$$\begin{aligned} a \cos \theta + b \sin \theta &= r\left(\frac{a}{r} \cos \theta + \frac{b}{r} \sin \theta \right) \\ &= r(\cos \alpha \cos \theta + \sin \alpha \sin \theta) \\ &= r \cos(\theta - \alpha) \;. \end{aligned}$$

Similarly, letting $\sqrt{A^2 + B^2} = R$, and $A/R = \cos 2\beta$, $B/R = \sin 2\beta$ we find

$$\begin{aligned} A \cos 2\theta + B \sin 2\theta &= R\left(\frac{A}{R} \cos 2\theta + \frac{B}{R} \sin 2\theta \right) \\ &= R(\cos 2\beta \cos 2\theta + \sin 2\beta \sin 2\theta) \\ &= R \cos 2(\theta - \beta) \;. \end{aligned}$$

Now f may be written in the form

$$f(\theta) = 1 - r \cos(\theta - \alpha) - R \cos 2(\theta - \beta) \;. \tag{1}$$

For $\theta = \alpha + 45°$ and $\theta = \alpha - 45°$, (1) yields

$$f(\alpha + 45°) = 1 - r/\sqrt{2} - R \cos 2(\alpha - \beta + 45°) \tag{2}$$

and

$$f(\alpha - 45°) = 1 - r/\sqrt{2} - R \cos 2(\alpha - \beta - 45°) \;. \tag{3}$$

If $r > \sqrt{2}$, then $1 - r/\sqrt{2} < 0$; and since the angles $2(\alpha - \beta) + 90°$ and $2(\alpha - \beta) - 90°$ differ by 180°, their cosines have opposite signs. Therefore one of the expressions

$$R \cos 2(\alpha - \beta + 45°), \qquad R \cos 2(\alpha - \beta - 45°)$$

is positive, so that the right side of one of the equations (2), (3) is negative. This means that at least one of the values $f(\alpha + 45°)$, $f(\alpha - 45°)$ is negative, contrary to the hypothesis. We conclude that $r^2 = a^2 + b^2 \leqslant 2$.

Similarly, we evaluate f at β and at $\beta + \pi$:

$$(4) \quad f(\beta) = 1 - r\cos(\beta - \alpha) - R\cos 0 = 1 - r\cos(\beta - \alpha) - R$$

and

$$(5) \quad \begin{aligned} f(\beta + \pi) &= 1 - r\cos(\beta - \alpha + \pi) - R\cos 2\pi \\ &= 1 - r\cos(\beta - \alpha + \pi) - R\,. \end{aligned}$$

If $R > 1$, then $1 - R < 0$; and since $\beta - \alpha$ and $\beta - \alpha + \pi$ differ by π their cosines have opposite signs. It follows that the right side of one of the equations (4), (5) is negative, i.e. that at least one of the values $f(\beta)$, $f(\beta + \pi) < 0$, contrary to the hypothesis. Hence $R^2 = A^2 + B^2 \leqslant 1$.

Remark. This is a special case of the theorem: If $f(\theta) = 1 + \Sigma_1^n(a_j \cos j\theta + b_j \sin j\theta) \geqslant 0$, then $a_j^2 + b_j^2 \leqslant 2$ for $j = 1, 2, \cdots, n-1$ and $a_n^2 + b_n^2 \leqslant 1$. This is an important property of the Fourier series of positive functions.

1977/5. According to the given information

$$(1) \quad a^2 + b^2 = q(a + b) + r \qquad \text{with} \qquad 0 \leqslant r < a + b\,,$$

and

$$(2) \quad q^2 + r = 1977\,.$$

These imply

$$(3) \quad q^2 \leqslant 1977 = q^2 + r < q^2 + a + b$$

and

$$(4) \quad a^2 + b^2 < (q + 1)(a + b)\,.$$

Since $2ab \leqslant a^2 + b^2$, (4) also implies that

$$(5) \quad 2ab < (q + 1)(a + b)\,.$$

Adding inequalities (4) and (5), we obtain

$$(a + b)^2 < 2(q + 1)(a + b)\,,$$

and on division by $(a + b)$, $a + b < 2q + 2$; so $a + b \leqslant 2q + 1$. Replacing $a + b$ in (3) by $2q + 1$ yields

$$q^2 \leqslant 1977 < q^2 + 2q + 1 = (q + 1)^2 .$$

The only integer q satisfying these inequalities is $q = 44$ because

$$44^2 = 1936 \qquad \text{and} \qquad 45^2 = 2025 .$$

Thus $q = 44$ and so $r = 41$ from (2) above. Substituting these values into (1), we find that

$$a^2 + b^2 = 44a + 44b + 41$$

or

$$(a - 22)^2 + (b - 22)^2 = 1009 .$$

By listing all squares up to 505 and their differences from 1009 (remembering that a square ends in 0, 1, 4, 9, 6, 5), we find that the only representation of 1009 as sum of two squares is

$$(15)^2 + (28)^2 = 1009 .$$

Therefore the sets $\{|a - 22|, |b - 22|\}$ and $\{15, 28\}$ must be the same. We conclude that $(a, b) = (50, 37), (37, 50), (50, 7)$, or $(7, 50)$.

1977/6 First solution. Note that f has a unique minimum at $n = 1$. For, if $n > 1$, we have

$$f(n) > f(f(n - 1)) .$$

The same reasoning shows that the next smallest value is $f(2)$, etc. Thus

$$f(1) < f(2) < f(3) < \dots .$$

Since $f(n) \geqslant 1$ for all n, we have, in particular, $f(n) \geqslant n$. Suppose that, for some positive integer k, $f(k) > k$. Then $f(k) \geqslant k + 1$; and since f is an increasing function, $f(f(k)) \geqslant f(k + 1)$, contradicting the given inequality. Therefore $f(n) = n$ for all n.

Second solution (due to Peter Schor). We shall establish the result

$$f(n) = n \tag{1}$$

by induction on n. Our proof of (1) for $n = 1$ consists of showing that

(a) $f(k) = 1$ for some k,

(b) if $f(k) = 1$ then $k = 1$.

Proof of (a). Suppose on the contrary, that $f(k) \neq 1$ for all k. Then we can construct the sequence

$$\begin{aligned} f(n + 1) > f(f(n)) &= f(n_1) > f(f(n_1 - 1)) \\ &= f(n_2) > \cdots = f(n_j) > f(f(n_j - 1)) > \cdots , \end{aligned} \tag{2}$$

where we have set $f(n) = n_1$, $f(n_1 - 1) = n_2, \cdots, f(n_{j-1} - 1) = n_j, \cdots$. Since the integers n_j are all greater than 1, $f(n_j - 1)$ is defined for all n_j and the sequence (2) has an infinite number of decreasing positive integer terms, which is clearly impossible. We conclude that 1 is in the range of f.

Proof of (b). By the given properties of f,

$$f(k) > f(f(k-1)) > 0$$

from which we conclude that

$$f(k) > 1 \qquad \text{for all } k > 1 . \tag{3}$$

This implies (b) and completes the first step of the induction, i.e. the proof of $f(1) = 1$.

To prove the validity of (1) for all n, we introduce an auxiliary function g defined by

$$g(n) = f(n+1) - 1 \tag{4}$$

and show first that g has the properties required of f by the problem, i.e. that

$$g(n+1) > g(g(n)) \quad \text{and} \quad g(n) > 0 \qquad \text{for } n = 1, 2, \ldots . \tag{5}$$

It follows from (3) that $g(n) = f(n+1) - 1 > 0$ for all n. To see that $g(n+1) > g(g(n))$, we use the definition (4) of g thrice:

$$g(g(n)) = g(f(n+1) - 1) = f(f(n+1)) - 1 ;$$

since $f(f(n+1)) < f(n+2)$ and by (4) $f(n+2) = g(n+1) + 1$, we have

$$g(g(n)) < f(n+2) - 1 = g(n+1) .$$

Since g satisfies properties (5), and since only these properties of f were used to prove that $f(1) = 1$, it follows that $g(1) = 1$. Now suppose that $g(k) = k$ for $k = 1, 2, \ldots, n-1$. From $g(n-1) = n-1$ it follows by definition (4) that $f(n) - 1 = n - 1$, whence $f(n) = n$. This completes the induction.

Note. Instead of pulling the function g out of a hat, we might have constructed it by the following rational procedure: We rewrite the condition

$$f(n+1) > f(f(n)) \tag{6}$$

given in the problem as

$$f \circ t(n) > f \circ f(n) \tag{6$'$}$$

where $\circ$ denotes composition of functions and t is the function

$$t(n) = n + 1 \tag{7}$$

Both, $t(n)$ and its inverse $t^{-1}(n) = n - 1$ are increasing functions. Now we apply (6)′ to the integer $t(n)$ instead of n getting

(6)″ $$f \circ t \circ t(n) > f \circ f \circ t(n) .$$

Since t^{-1} is monotonically increasing, applying it on the left of (6)″ preserves the inequality:

$$t^{-1} f \circ t \circ t(n) > t^{-1} \circ f \circ f \circ t(n) .$$

Finally, we insert the identity $t \circ t^{-1}$ between f and f on the right to obtain

(6)‴ $$(t^{-1} \circ f \circ t) \circ t(n) > (t^{-1} \circ f \circ t) \circ t^{-1} \circ f \circ t(n) ,$$

where the function $g(n) = t^{-1} \circ f \circ t(n) = f(n + 1) - 1$ emerges†, just as defined by (4). In terms of g (9)‴ states

$$g \circ t(n) > g \circ g(n) , \quad \text{i.e.} \quad g(n + 1) > g(g(n)) .$$

†The functions f and $t^{-1} \circ f \circ t$ can be interpreted as the same mappings in different coordinate systems.

List of Symbols

(ABC)	area of $\triangle ABC$
$\simeq$	approximately equal to
$\cong$	congruent (in geometry)
$a \equiv b (\mathrm{mod}\, p)$	$a - b$ is divisible by p; *Congruence*, see glossary
$a \not\equiv b (\mathrm{mod}\, p)$	$a - b$ is not divisible by p
$\equiv$	identically equal to
$[x]$	integer part of x, i.e. greatest integer not exceeding x
$\binom{n}{k}$, $C(n, k)$	binomial coefficient, see glossary; also the number of combinations of n things, k at a time
(n, k), G.C.D. of n, k	the greatest common divisor of n and k
$p \mid n$	p divides n
$p \nmid n$	p does not divide n
$n!$	n factorial $= 1 \cdot 2 \cdot 3 \cdot \ldots (n-1)n, \quad 0! = 1$
$\prod_{i=1}^{n} a_i$	the product $a_1 \cdot a_2 \cdot \ldots \cdot a_n$
$\sim$	similar in geometry
$\sum_{i=1}^{n} a_i$	the sum $a_1 + a_2 + \ldots + a_n$
$\circ$	$f \circ g(x) = f[g(x)]$, see *Composition* in glossary
$K_1 \cup K_2$	union of sets K_1, K_2
$K_1 \cap K_2$	intersection of sets K_1, K_2
A. M.	arithmetic mean, see *Mean* in glossary
G. M.	geometric mean, see *Mean* in glossary
H. M	harmonic mean, see *Mean* in glossary
$[a, b]$	closed interval, i.e all x such that $a \leqslant x \leqslant b$
(a, b)	open interval, i.e. all x such that $a < x < b$

Glossary of some frequently used terms and theorems.

Arithmetic mean (average). see *Mean*

Arithmetic mean-geometric mean inequality.

If $a_1, a_2, \ldots, a_n$ are n positive numbers, then

(1) $\frac{1}{n} \sum_{i=1}^{n} a_i \geqslant \left(\prod_{i=1}^{n} a_i \right)^{1/2}$;= holds if and only if $a_1 = a_2 = \ldots = a_n$.

Proof for $n = 2$:

$$(\sqrt{a_1} - \sqrt{a_2})^2 \geqslant 0 \Leftrightarrow a_1 + a_2 \geqslant 2\sqrt{a_1 a_2} \Leftrightarrow \frac{a_1 + a_2}{2} \geqslant \sqrt{a_1 a_2} .$$

(For elementary proofs valid for all n, see e.g. the *Hungarian Problem Book* I, NML vol. 11, 1963, p. 70.)

Arithmetic mean-harmonic mean inequality

(2) $\frac{a+b}{2} \geqslant \frac{2ab}{a+b} = \left[\frac{1}{2} \left(\frac{1}{a} + \frac{1}{b} \right) \right]^{-1}$ for $a, b > 0$; = holds if and only if $a = b$.

Proof: $\frac{(a-b)^2}{2(a+b)} \geqslant 0 \Leftrightarrow \frac{(a+b)^2 - 4ab}{2(a+b)} \geqslant 0 \Leftrightarrow \frac{a+b}{2} - \frac{2ab}{a+b} \geqslant 0.$

Arithmetic series. see *Series*

Binomial coefficient:

$$\binom{n}{k} = \frac{n!}{k!(n-k)!} = \text{coefficient of } y^k \text{ in the expansion } (1 + y)^n$$

(See Binomial theorem below and List of Symbols.)

Binomial theorem:

$$(x+y)^n = \sum_{k=0}^{n} \binom{n}{k} x^{n-k} y^k, \quad \text{where}$$

$$\binom{n}{k} = \frac{n(n-1)\dots(n-k+1)}{1\cdot 2 \dots (k-1)k} = \frac{n!}{k!(n-k)!}\,.$$

Cauchy's theorem. stated and proved in 1975/6.

Centroid of $\triangle ABC$.
Point of intersection of the medians

Centroid
Center of gravity

Characteristic equation of a matrix M. defined in 1963/4.
$\det(M - \lambda I) = 0$

Characteristic value of a matrix M.
Solution of characteristic equation; see *Eigenvalue*

Characteristic vector of a matrix M. see *Eigenvector*

Chebyshev polynomial.
Polynomial $T_n(x)$ expressing $\cos n\theta$ as a polynomial in $x = \cos\theta$.

Circumcenter of $\triangle ABC$.
Center of circumscribed circle of $\triangle ABC$

Circumcircle of $\triangle ABC$.
Circumscribed circle of $\triangle ABC$

Complex numbers.
Numbers of the form $x + iy$, where x, y are real and $i = \sqrt{-1}$.

Composition of functions.
$F(x) = f\circ g(x) = f[g(x)]$ is the composite of functions f, g, where the range of g is the domain of f.

Congruence.
$a \equiv b \pmod p$ read "a is congruent to b modulo p" means that $a - b$ is divisible by p.

Convex function. defined in 1969/6.
It follows from the definition that if $f(x)$ is continuous, and if

$$f(a) + f(b) \geqslant 2f\left(\frac{a+b}{2}\right) \text{ for all } a, b \text{ in } I\,,$$

then $f(x)$ is convex. If $f(x)$ is twice differentiable in I, then $f(x)$ is convex if and only if $f(x) \geqslant 0$ in I. The graph of a convex function lies above its tangent. See also Jensen's theorem, below.

Convex hull of a pointset S.
The intersection of all convex sets containing S.

Convex pointset.
A pointset S is convex if, for every pair of points P, Q in S, all points of the segment PQ are in S.

Construction of locus of points X *such that* $\angle AXB$ *has a given measure,* A, B *given points.* explained in 1961/5

Cyclic quadrilateral.
Quadrilateral that can be inscribed in a circle.

De Moivre's theorem.
$(\cos\theta + i\sin\theta)^n = \cos n\theta + i\sin n\theta$.
For a proof, see 1962/4, p. 49

Determinant of a square matrix M (det M):
A multi-linear function $f(C_1, C_2, \ldots, C_n)$ of the columns of M with the properties

$$f(C_1, C_2, \ldots C_i \ldots C_j \ldots C_n) = -f(C_1, C_2 \ldots C_j, \ldots C_i, \ldots C_n)$$

and det $I = 1$.

Difference equations.
Linear difference equations: See Recursions

Dirichlet's principle. See Pigeonhole principle

Dot product (*scalar product*) $V_1 \cdot V_2$ *of two vectors.*
If $V_1 = (x_1, y_1, z_1)$, $V_2 = (x_2, y_2, z_2)$, then their dot product is the number $V_1 \cdot V_2 = x_1x_2 + y_1y_2 + z_1z_2$. If θ is the angle between V_1 and V_2 and $|V|$ denotes the length of V, then $V_1 \cdot V_2 = |V_1||V_2| \cos\theta$.

Eigenvalue (*characteristic value*) of a matrix M. defined in 1963/4.

Eigenvector (*characteristic vector*) of a matrix M. defined in 1963/4.

Euclid's algorithm.
A process of repeated divisions yielding the greatest common divisor of two integers, $m > n$:

$$m = nq_1 + r_1 , \quad q_1 = r_1 q_2 + r_2 , \ldots q_k = r_k q_{k+1} + r_{k+1} ;$$

the last non-zero remainder is the GCD of m and n. (See e.g. *Continued Fractions* by C. D. Olds, NML vol. 9, 1963, p. 16)

Euler's theorem on the distance d between in- and circumcenters of a triangle.

$$d = \sqrt{R^2 - 2rR} ,$$

where r, R are the radii of the inscribed and circumscribed circles.

Euler-Fermat theorem. stated and proved in 1971/3

Euler's function $\phi(p)$. defined in 1971/3.

Excircle of $\triangle ABC$.
Escribed circle of $\triangle ABC$

Fermat's theorem. stated and proved in 1970/4.

Geometric mean. see *Mean*, geometric

Geometric series. see *Series*, geometric

Harmonic mean. see *Mean, harmonic*

Harmonic mean-geometric mean inequality.
For $a, b > 0$,

$$(3) \quad \frac{2ab}{a+b} \leqslant \sqrt{ab} ; = \text{ holds if and only if } a = b.$$

Proof: From arithmetic mean-geometric mean inequality $(a + b)/2 \geqslant \sqrt{ab}$, see (1), it follows that $2/(a + b) \leqslant 1/\sqrt{ab}$. Multiply by ab to obtain (3).

Heron's formula.
The area (ABC) of $\triangle ABC$ with sides a, b, c is

$$(ABC) = [s(s-a)(s-b)(s-c)]^{1/2} , \quad \text{where } s = \frac{1}{2}(a + b + c) .$$

Homogeneous.
$f(x, y, z, \ldots)$ is *homogeneous of degree k* if

$$f(tx, ty, tz, \ldots) = t^k f(x, y, x, \ldots) .$$

A system of linear equations is called homogeneous if each equation is of the form $f(x, y, z, \quad) = 0$ with f homogeneous of degree 1.

Incenter of $\triangle ABC$.
Center of inscribed circle of $\triangle ABC$

Incircle of $\triangle ABC$.
Inscribed circle of $\triangle ABC$

Inequalities.

Arithmetic mean-geometric mean-. see *Arithmetic mean*

Arithmetic mean-harmonic mean-. see *Arithmetic mean*

Cauchy-Schwarz-Buniakowski-.

$|X \cdot Y| \leqslant |X||Y|$ or

$$|x_1y_1 + x_2y_2 + \dots + x_ny_n| \leqslant \left[\sum x_i^2\right]^{1/2}\left[\sum y_i^2\right]^{1/2};$$

see 1970/5 for a proof. This inequality was first proved by Cauchy for finite dimensional vector spaces, then generalized by Buniakowski and independently by Schwarz.

Harmonic mean-geometric mean-. see *Harmonic mean*

Schwarz-. see *Cauchy-Schwarz-Buniakowski*

Triangle-.

$|X + Y| \leqslant |X| + |Y|$.

Inverse function.
$f: X \to Y$ has an inverse f^{-1} if for every y in the range of f there is a unique x in the domain of f such that $f(x) = y$; then $f^{-1}(y) = x$, and $f^{-1}\circ f$, $f \circ f^{-1}$ are the identity functions.

Isoperimetric theorem for triangles.
Among all triangles with given area, the equilateral triangle has the smallest perimeter. (See reference given in footnote, p. 38)

Jensen's theorem.
If $f(x)$ is convex in an interval I then

$$f(a_1) + f(a_2) + \dots + f(a_n) \geqslant nf\left(\frac{a_1 + a_2 + \dots + a_n}{n}\right)$$

for all a_i in I.

Matrix.
A rectangular array of numbers (a_{ij})

Mean of n numbers.

Arithmetic mean (average) $= \frac{1}{n}\sum_{i=1}^{n} a_i$

Geometric mean $= \sqrt[n]{a_1a_2\dots a_n}$, $a_i \geqslant 0$

Harmonic mean $= \left(\frac{1}{n}\sum_{i=1}^{n}\frac{1}{a_i}\right)^{-1}$, $a_i > 0$

Orthocenter of $\triangle ABC$.
Intersection of altitudes of $\triangle ABC$

Periodic function.
$f(x)$ is periodic with period a if $f(x+a)=f(x)$ for all x.

Pigeonhole principle (*Dirichlet's box principle*).
If n objects are distributed among $k<n$ boxes, some box contains at least two objects.

Polynomial in x *of degree* n:
Function of the form $P(x)=\sum_{i=0}^{n} c_i x^i, \quad c_n \neq 0.$

Prismatoid. defined in 1965/3

Prismatoid formula. see 1965/3 for the formula and two proofs of its validity.

Quadratic form.
$q=\sum_{i,j=1}^{n} c_{ij}x_i x_j$, i.e. a homogeneous function of degree 2 (see *Homogeneous*). Also written $q(X)=X\cdot CX$ with $X=(x_1, x_2 \ldots x_n)$, and C the symmetric matrix (c_{ij}).

Quadratic residue (mod p). defined in 1970/4.

Quartic form.
homogeneous function of degree 4.

Ramsey's theorem. stated in 1964/4

Recursion.
Linear recursions, matrix treatment in 1963/4, 1974/3.

Root of an equation.
Solution of an equation

Roots of a polynomial related to its coefficients.
Relations are obtained by equating coefficients of like powers of x in the identity

$$P(x)=\sum_{i=1}^{n} c_i x^i \equiv c_n(x-x_1)(x-x_2)\ldots(x-x_n),$$

where x_i are the zeros of $P(x)$.

Series.

Arithmetic: $\sum_{j=1}^{n} a_j$ with $a_{j+1} = a_j + d$, d the *common difference.*

Sum of —: $\sum_{j=1}^{n} a_j = \frac{n}{2}[2a_1 + (n-1)d]$.

Geometric: $\sum_{j=0}^{n-1} a_j$ with $a_{j+1} = ra_j$, r the *common ratio.*

Sum of —: $S_n = \sum_{j=0}^{n-1} a_0 r^j = a_0 \sum_{j=0}^{n-1} r^j = a_0 \frac{1-r^n}{1-r}$.

Proof: $S_n - rS_n = a_0 \sum_{j=0}^{n-1} (r^j - r^{j+1})$

$= a_0[r^0 - r^1 + r^1 - r^2 + \ldots + r^{n-1} - r^n]$,

where the sum in brackets "telescopes" to $1 - r^n$. Thus $S_n(1-r) = a_0(1-r^n)$, from which above follows.

Sum S of infinite geometric series for $|r| < 1$:

$$\sum_{j=0}^{\infty} a_0 r^j = \lim_{n\to\infty} \sum_{j=0}^{n} a_0 r^j = \lim_{n\to\infty} a_0 \frac{1-r^n}{1-r} = \frac{a_0}{1-r} .$$

Stewart's theorem. stated in 1960/3.

Sum of series. see *Series*

Telescoping sum.

$\sum_{1}^{n} (a_i - a_{i+1}) = a_1 - a_2 + a_2 - a_3 + \ldots + a_{n-1} - a_n = a_1 - a_n$,

i.e. a sum whose middle drops out through cancellation, see 1966/4. 1970/3 for examples.

Triangle inequality. see *Inequalities*

Vectors.

ordered n-tuples obeying certain rules of addition and multiplication by numbers.

Vector representation of points, segments. briefly explained in 1960/5, used subsequently.

Wilson's theorem. stated and proved in 1970/4.

Zero of a function $f(x)$.

any point x for which $f(x) = 0$.

References

General:

1. W. W. R. Ball and H. S. M. Coxeter, *Mathematical Recreations*, Macmillan, 1939.
2. R. Courant and H. Robbins, *What is Mathematics*? Oxford University Press, 1941.
3. R. Honsberger, *Mathematical Gems*, Vol. I, II., M.A.A. 1973, 1976.
4. G. Pólya, *How To Solve It*, Princeton University Press, 1948.
5. G. Pólya, *Mathematics and Plausible Reasoning*, Vol. I, Princeton University Press, 1954.

Algebra:

6. S. Barnard and J. M. Child, *Higher Algebra*, Macmillan Co., 1939.
7. Hall and Knight, *Higher Algebra*, Macmillan Co., 1964.

Geometry:

8. N. Altshiller-Court, *College Geometry*, Barnes and Noble, 1957.
9. N. Altshiller-Court, *Modern Pure Solid Geometry*, Chelsea Press, 1964.

Others:

10. R. D. Carmichael, *The Theory of Numbers and Diophantine Equations*, Dover Publications, 1959.
11. S. Goldberg, *Introduction to Difference Equations*, John Wiley and Sons, 1958.
12. D. S. Mitrinovic, *Elementary Inequalities*, P. Noordhoff, Ltd., Gröningen, Netherlands, 1964.
13. J. Riordan, *Introduction to Combinatorial Analysis*, John Wiley and Sons, 1958.

Problems:

14. G. Pólya and G. Szegö, *Problems and Theorems in Analysis*, Vol. I, II, Springer-Verlag, 1972.
15. D. O. Shklarsky, N. N. Chentzov and I. M. Yaglom, *The USSR Olympiad Problem Book*, W. H. Freeman, 1962.
16. A. M. Yaglom and I. M. Yaglom, *Challenging Mathematical Problems with Elementary Solutions*, Holden-Day, 1964.

Also:

The entire collection of the New Mathematical Library (now 27 volumes–see p. vi of this book) available from the Mathematical Association of America is highly recommended.